Intersectional Lives

Intersectional Lives explores the varied experiences of Chinese Australian women across time and place during the White Australia Policy era (1901–1973). Chinese Australian women's personal reflections are examined alongside postcolonial feminist readings of official records to illustrate how their everyday lives were influenced by multiple and fluid identities and subject positions including migrant, mother, daughter, wife, student, worker, entrepreneur and cultural custodian. This book provides new ways to conceptualise Chinese females in the diaspora as gendered, classed, culturally varied and racialised individuals with multiple forms of oppression, agency and mobility. It offers a revision of patriarchal understandings of Chinese Australian history and broader understandings of overseas Chinese migrations and settlement experiences. It also demonstrates how historical geography, informed by postcolonial feminist approaches, can facilitate more nuanced understandings of past (and present) times and places that include women's diverse experiences at the domestic, local, national and international scale.

This book will appeal to social and cultural geographers with additional audiences of interest in history and historical geography, ethnic and racial studies, gender studies, diaspora studies, migration studies, and gender and feminist studies.

Alanna Kamp is Lecturer in Geography and Urban Studies at Western Sydney University. Her research contributions lie in the areas of Australian multiculturalism and cultural diversity, experiences of migration and migrant settlement, racism, national identity and intersectional experiences of belonging/exclusion. She has published pioneering and award-winning works on Chinese Australian women in White Australia.

Routledge International Studies of Women and Place

Series Editors: Janet Henshall Momsen, *University of California, Davis*

Who Will Mind the Baby?
Geographies of Childcare and Working Mothers
Edited by Kim England

Feminist Political Ecology
Global Issues and Local Experience
Edited by Dianne Rocheleau, Barbara Thomas-Slayter and Esther Wangari

Women Divided
Gender, Religion and Politics in Northern Ireland
Rosemary Sales

Women's Lifeworlds
Women's Narratives on Shaping their Realities
Edited by Edith Sizoo

Gender, Planning and Human Rights
Edited by Tovi Fenster

Gender, Ethnicity and Place
Women and Identity in Guyana
Linda Peake and D. Alissa Trotz

Brokering Circular Labour Migration
A Mobile Ethnography of Migrant Care Workers' Journey to Switzerland
Huey Shy Chau

Climate Change, Gender Roles and Hierarchies
Socioeconomic Transformation in an Ethnic Minority Community in Vietnam
Phuong Ha Pham and Donna L. Doane

Intersectional Lives
Chinese Australian Women in White Australia
Alanna Kamp

Intersectional Lives

Chinese Australian Women in White Australia

Alanna Kamp

Routledge
Taylor & Francis Group

LONDON AND NEW YORK

First published 2022
by Routledge
4 Park Square, Milton Park, Abingdon, Oxon OX14 4RN

and by Routledge
605 Third Avenue, New York, NY 10158

Routledge is an imprint of the Taylor & Francis Group, an informa business

British Library Cataloguing-in-Publication Data
A catalogue record for this book is available from the British Library

Library of Congress Cataloging-in-Publication Data
A catalog record has been requested for this book

ISBN: 978-0-367-67429-8 (hbk)
ISBN: 978-0-367-67430-4 (pbk)
ISBN: 978-1-003-13133-5 (ebk)

DOI: 10.4324/9781003131335

Typeset in Goudy
by SPi Technologies India Pvt Ltd (Straive)

Contents

List of figures   vi
List of tables   vii
Acknowledgements   viii
A Note on Terminology   x
A Note on Historical Photographs   xi

1 Introduction   1

2 Intersectionality and postcolonial feminist geography
   as a way of inclusion   25

3 Presence, diversity and mobility   51

4 Domestic roles and the family economy   87

5 Cultural maintenance in homes and families   115

6 Interactions and identity   144

7 Conclusions and looking forward   169

   Appendix A: Brief biographies of interview participants   183

   Appendix B: List of census publications   194

   Index   200

Figures

3.1   Mei-Lin Yum and cousin Coral in Texas, Queensland, c.1950.   62

3.2   Mabel Lee and her sisters in Merrylands, Sydney c.1945.   63

3.3   Two generations of the Lumbiew family, Sydney, c.1900.   65

3.4   Four generations of the Fay family, 1941.   66

3.5   The extended Fay family, c. 1940.   67

3.6   The Loong family, Melbourne, c. 1917.   68

4.1   Doreen Cheong née Lee and her brothers, c.1947.   90

4.2   Nancy Buggy née Loong with her siblings, Parramatta, Sydney, c.1945.   93

4.3   Mei-Lin Yum and her parents at the Prefects Induction, Ballina 1960.   95

4.4   Fong See Lee Yan and first-born daughter outside her and her husband's long and short soup shop in Cairns, Queensland, in the early 1940s.   103

4.5   Sally Pang with family members, 1969.   104

4.6   Staff of the Hong Yuen lady's showroom, Inverell, 1949.   105

5.1   Dressed in traditional Chinese costume for the Empire Day Parade, Texas, Queensland, c. 1945.   134

6.1   Patricia Yip and her school friends, Sydney, New South Wales, c. 1955.   158

6.2   Doreen Cheong celebrating a birthday, Ballina, New South Wales, c. 1955.   163

Tables

3.1 'Full Chinese' and 'Mixed Chinese' in Australia, 1901–1971   52
3.2 Australian-Born and Foreign-Born Chinese Females in Australia, 1911–1966   53
3.3 Country of Birth of Foreign-Born Chinese Australian Females, 1911–1961   55
3.4 Marriage Status of 'Full Chinese' and 'Mixed Chinese' Females in Australia, 1911–1961   58
3.5 Age Groups of Chinese Australian Females, 1911–1966   62
3.6 Length of Female Chinese Migrant Residence in Australia, 1911–1961   64
3.7a Chinese Arrivals and Departures to/from Australia, 1914–1947   70
3.7b Chinese Arrivals and Departures to/from Australia, 1949–1965   71
3.8 State Distributions of Chinese Australian Females, 1911–1966   78
3.9 Rural/Urban Distributions of Female Chinese Australians, 1911–1966   80
4.1 Chinese Australian Female Dependants in Australia, 1911   107
4.2 Occupations of Chinese Australian Females, 1911   108
4.3 Employment of Chinese Australian Females, 1921 and 1933   109
5.1 Literacy of Chinese Australian Females, 1911   119
A.1 Interview Participant Information   183

Acknowledgements

Intersectional Lives would not exist without the contributions of the 19 women who participated in my research project. These women enthusiastically shared their memories, experiences and family histories with me, often welcomed me into their homes and gave me a glimpse into their lives. My gratitude for their contributions is immeasurable.

This book began its life as a PhD thesis completed at Western Sydney University in 2014. The research was funded by an Australian Postgraduate Award and supported by the School of Social Sciences, Western Sydney University. Western Sydney University occupies the traditional lands of the Darug, Eora, Dharawal (also referred to as Tharawal) and Wiradjuri peoples. I also acknowledge that my research (and writing of this book) took place on these unceded lands. I thank the Darug, Eora, Dharawal (Tharawal) and Wiradjuri elders and their communities for their support of the university's work on their lands.

I am greatly indebted to Professor Kevin Dunn and Professor Emma Waterton who supervised me through the journey of a PhD. Now, as mentors and colleagues, they have continued their support of my work, encouragement of this project's conversion into a monograph, and willingness to read and comment on drafts.

I am also thankful to my PhD thesis examiners, Professor Louise Johnson and Professor Mona Domosh. Their enthusiasm for this project many moons ago and encouragement to get this work published has kept me inspired and determined. Special thanks must go to Professor Domosh who provided many insightful suggestions of how this project can be extended. I have included her recommendations in the section 'Directions for future research' in the concluding chapter of this book. I also thank Professor Janice Monk and Professor Janet Momsen, the editors of the series in which this book is published, for their support of *Intersectional Lives*. More broadly, the feminist contributions of these four pioneering geographers—Janice Monk, Janet Momsen, Mona Domosh and Louise Johnson—to the academy have provided much grounding for this project. To follow in their footsteps is a privilege and honour.

This book is also derived in part from the following published journal articles and book chapter:

Kamp, A. 2013, 'Chinese Australian Women in White Australia: Utilising Available Sources to Overcome the Challenge of "Invisibility"', *Chinese Southern Diaspora Studies*, no.6. (Copyright Alanna Kamp).

Kamp, A. 2018, 'Chinese Australian women's 'homemaking' and contributions to the family economy in White Australia', *Australian Geographer*, vol. 49, no. 1, pp. 149–165. (Copyright the Geographical Society of NSW).

Kamp, A. 2021, 'Chinese Australian Daughters' Experiences of Educational Opportunity in 1930s–60s Australia', *Australian Historical Studies*, pp. 1–18 (online). (Copyright Editorial Board, Australian Historical Studies).

Kamp, A. 2021, 'International migration and mobility experiences of Chinese Australian women in White Australia, 1901–1973', in K. Bagnall and J. Martínez (eds.), *Locating Chinese Women: Historical Mobility between China and Australia*, HKU Press, Hong Kong, pp. 105–128. (Copyright Hong Kong University Press).

Where necessary, I have noted these prior publications in the relevant chapters of this book. The contributions of editors and anonymous peer-reviewers in the publication of these articles and chapter have also helped craft this book.

My acknowledgements would not be complete without thanking my family and friends who have supported me throughout this journey. Their help and encouragement have been a constant. Special thanks must go to my partner Lewis and my parents whose own migration stories and experiences have no doubt inspired my work.

A Note on Terminology

The majority of Chinese names (of people and places) used in this book were most often provided to me by interview participants. Sometimes Romanised versions of Chinese names were used, for example, participants often referred to Guangdong province (rather than the anglicised form of Canton) and used the Romanised versions of family members' Chinese names such as Fong See Lee Yan. Other times, anglicised forms of Chinese names or English names were used; for example, in her interview, Helen Sham-Ho used her anglicised name rather than her Chinese name, Sham-Ho Wai-Har. I have not attempted to standardise names of people and places but have, instead, maintained the use and spelling of the terms/names as they were provided to me by the interview participants.

I use the term 'Chinese Australian' throughout this book in reference to Chinese immigrants to Australia and their Australian-born descendants who identify/identified themselves as having Chinese ancestry/being 'Chinese' (but not necessarily 'Chinese Australian'). My use of 'Chinese Australian' does not denote citizenship status and does not reflect the exclusionary official (government) terminology of the time. Chinese immigrants were unable to apply for naturalisation until 1956 (and then only under various provisions). The use of the term 'Chinese Australian' is therefore used to acknowledge the ethnic/cultural identity, presence and contributions of these individuals in the Australian nation.

A Note on Historical Photographs

The family photographs and portraits presented in this book were volunteered to me by research participants. These photographs are historical in nature (dating back to c.1900) and are privately owned/in private collections rather than institutional repositories. All attempts have been made to clear permissions. Please send any queries to the publisher.

1 Introduction

My grandmother came out to be the concubine for a general merchant and pearl dealer and she had no background material of her own. She had been sold as a baby. That was all she ever knew. And she and some other girls, babies, were adopted by a woman who raised them and married them off. Then she brought out a couple, my grandmother included, and married her off to my grandfather.

My mother was born in 1906 on Thursday Island. She was the first of nine children. I never knew my father. We had a photograph of him and my mother being married in Hong Kong. He was in a long black gown, sitting on a chair and she had the usual pose behind him with her hand on the chair, looking very sad. She didn't get married until she was about twenty-one and they decided that she was getting really old. So they arranged for her to go to Hong Kong and marry a friend's son. She was very unhappy apparently, but they had a big wedding in the Hong Kong Hotel. She said if she had to get married, well she was going to have a nice big wedding.

She had originally gone back to Hong Kong to take her father over to do what the usual thing was—to die in his village. He was very sick. And so that was her story. She left her husband in Hong Kong to come back to Thursday Island to have me after her first baby died. When she went back to find him, my father, he was gone and she never found out where he went to.

I think at the age of seven or eight months she decided to leave me on Thursday Island while she went back to see what was going on with her marriage. But not being able to find him, she had to give up and when she came home, I went back to her for a little while, but then she met another man and had my first half-brother. She went back to Hong Kong with him because he was sick and he wanted some traditional Chinese medicine. And so she left me again with my grandmother. By that time, when she came back, I didn't want to go back to her. I was quite happy running around.

I started school early because I used to follow my uncle to school. I think I was only four and a half. Everyday I'd follow him up the road. It wasn't a very big island. It was very small. You could run three blocks up there and that was the school. And I'd wait 'til they'd go to school and follow them. So they said, 'she might as well stay'. So I started school early and I remember coming home when I was still in the early years of school. So I must've been

DOI: 10.4324/9781003131335-1

five or five and a half. And they said, 'Guess who's in the office?' So I went in to have a look. They said, 'Do you know who it is?' And I said, 'Yes', not having a clue. And it was my mother. So she was very upset because I didn't know her and didn't want her and I was scared that I'd have to go back to her.

She had a couple of 'associations', you might say. These days they're just boyfriends and she didn't marry until my youngest half-brother was born— just before he was born. The half-brothers who came before the last one, were actually born to...oh how shall I put it? Out of these alliances—two men she met, not both at the same time, but these were successive periods during her life. She had a very hard life and most of it was survival, I think. They were quite nice men, but one thing led to another. One of them died and one had stayed behind in Hong Kong when she went over again because of the war. She never saw him again.

There was a period when I didn't see her for quite some time. She had a business on Thursday Island and she lived down one end of the island running a shop and over the years, I think, she had a couple of babies too. I think my two half-brothers were born and my half-sister was born to this man who took her to Hong Kong. So she had this little group of children and I had an attraction. There was an attraction there. I used to get down and play with them but go home to my grandmother. My sister told me recently that she never even knew until they were in their teens who I was. They thought I was just some cousin who'd come over to play.

I didn't know my grandfather. My mother took him back to Hong Kong and then got married, so I wasn't even born, but having been brought up by my grandmother I got a lot of stories about him, a lot of which I have forgotten. And she didn't have any education herself. She was brought up as a servant more or less, hired out to work for people. As I said, she was sold when she was a baby, and so she was full of folk stories, some of which I remember, and gossip which she fed me. I was fed so much gossip because she didn't have anyone else to talk to and she couldn't speak much English either.

We spoke English because my grandmother always wanted to learn English. So she would take my kindergarten books and learn, and she would learn to read. She did learn to read, too. She learnt to write her name and she wanted to do so many things. It's really terrible, I think she must've been a really intelligent person, and she just didn't have that opportunity. That was her main grievance—that she had no education of any sort. But I still remember her, having a rest after lunch on the kitchen bench she lay on and she'd read my school books.

We chose Cairns because it was the closest and my grandmother wanted to high-tail it back to the island as soon as the war broke out, thinking it would stop next year. No one knew when it was going to stop. And she did that. We were four years in Cairns and as soon as we were allowed to, we took a little cargo boat back, a two cabin cargo boat, and the grass was thigh high on the island and bullet holes were in the walls. Everything had been ransacked and the only photo I ever had of my mother and father...the only photo I would ever have, we left on the wall. So, that's a shame. Well the soldiers just came in and I think they used the walls for target practice.

I only stayed there a year or two, about two years at the most, and decided to go back to go to my mother and I stayed with her a few months and took off to Sydney. I'd just turned seventeen.

(Interview with Stella Sun, 2010)

This story belongs to Stella Sun,[1] a Chinese Australian woman who was born on Thursday Island in 1931. The account is, in fact, a series of memories that Stella recounted to me during an interview I conducted with her in 2010. By piecing her memories together, Stella's family history on the island, which dates back to c.1900, tells a rich story of migration and mobility, settlement, family relationships, gender roles, isolation, identity, belonging, sacrifice, resistance and contribution within a very specific historical, cultural and geographical context. Thursday Island is a small island (3 km²) that lies off the coast of far north Queensland in the Torres Strait (a narrow waterway that connects the Coral and Arafura Seas between the northernmost tip of Australia and New Guinea). The Kaurareg people are the traditional owners of the area; however, by the nineteenth century, European settlers had established Thursday Island as an important stop for trading vessels and the commercial and administrative centre of the Torres Strait. Thursday Island has also played an important role in the Australian pearl industry. This was an industry which, from the nineteenth century, saw thousands of immigrants from diverse groups participate. These groups included peoples from Asia, the South Pacific and Europe. In World War II, Thursday Island was established as an important military base and today the economy is centred on traditional activities such as fishing. Stella's story is set between the 1930s and 1940s—a time in which Thursday Island was operating as a trading post, pearling centre, military base and was subsequently culturally diverse. Paradoxically, this period was also a time in which Australian immigration legislation and other social policies were framed by the broader White Australia Policy.

The White Australia Policy was an umbrella term used to refer to an array of legislative measures for non-European exclusion. Based on ideas of national self-determination and survival, the policy responded to a desire for a nation inhabited only by Whites[2] and dominated Australia's formal stance on immigration and Indigenous affairs from Federation of the nation in 1901 until the multicultural policies of the 1970s (Palfreeman 1967; Yarwood 1964; London 1970; Elder 2005; Walker 2005; and Fitzgerald 2007). The *Immigration Restriction Act 1901* was the first formal federal legislative tool of the White Australia Policy and aimed to protect the nation from the external Other (rather than the internal Indigenous Other) by prohibiting non-European immigration by means of an educational test in any European language. It is widely acknowledged that arguments for nationally unified and systematised immigration restriction in 1901 can be traced to nineteenth-century anti-Chinese sentiments and colonial legislation which severely restricted the entry of Chinese and other 'coloured races' (Palfreeman 1967; Elder 2005; and Kamp 2010). Despite these colonial restrictions and the consequent decrease in the number of non-White migrants entering Australia, fears of an influx of 'Asian hordes' which would threaten the development of a White Australia continued. This fuelled the insistence for national exclusionary measures throughout the twentieth century (Yarwood 1964; London 1970; and Walker 2005).

Reflecting on her life during this period of Australian history in which strident racism (among other forms of discrimination) were sanctioned by the state and impacted the lives of thousands of non-White Australians, Stella turned to me and asked in surprise, 'does this interest you?' Sitting in the lounge room of her modest home in north-western Sydney—almost 3,000 kilometres from Thursday Island—this question shocked me. Her story was so obviously captivating and was providing extraordinary personal insights into life as a Chinese Australian female[3] in the context of White Australia. I would soon come to realise that Stella's assumption that her life experiences were mundane was not out of the ordinary. Another interviewee, Susan, apologised to me during our interview exclaiming, 'I'm sorry. I think it's been terribly boring'. Additionally, many women I met during my research showed a keen interest in my project but were unwilling to participate in interviews because they believed their story was unimportant or uninteresting.

I have chosen to begin this book with Stella's recollections for two reasons. Firstly, because her story *is* important and I *do* find her experiences interesting (and I hope you do too). Secondly, and more importantly, I have chosen to open with Stella's voice because the driving objective of *Intersectional Lives* is to give Chinese Australian females a space to tell *their own stories*. These stories, told in their own words, tell of ways in which varied experiences as migrants, mothers, daughters, wives, students, workers and business owners in White Australia were influenced by multiple and fluid identities and subject positions within gendered, racialised and classed structures of power. Such intersectionality was at times oppressive and constraining—resulting in social, economic and political exclusions within homes, schools and universities, neighbourhoods, communities, workplaces and the nation. At other times, intersectional forces of oppression were negotiated, resisted or challenged so that contributions and experiences of inclusion were possible, defying official race-based exclusion. Therefore, it is through the research participants' voices that I argue throughout this book that the politics of belonging and exclusion in White Australia was intrinsically connected to such intersectionality.

Stella's voice is just one of the thousands of women's voices that are missing from scholarly accounts of Chinese experiences in the White Australia Policy period and more general examinations of Australian 'race' relations and nation-building. The historic immigration of Chinese to Australia and their settlement experiences have been of increasing interest to historians and other social researchers since the 1960s (Kamp 2013). Against a backdrop of anti-Chinese legislation, key considerations for researchers have included inter-racial relations; explanations of the administration and impacts of discriminatory legislation; White Australian attitudes towards Chinese; and political, economic and social challenges and successes of (male) Chinese Australians (see, for example, Yarwood 1964; Palfreeman 1967; London 1970; Choi 1975; and Yong 1977).

Since the 1980s, research has offered a counterpoint to this 'top down' approach, increasingly 'attempt[ing] an understanding of Chinese on their own terms' (Markus 1983: 89; see, for example, Giese 1997; Macgregor 1998; Tan 2003; Couchman 2004, 2011; Wilton 2004; and Bagnall 2006, 2011). Despite such research momentum, the lived experiences of Chinese Australians in the White Australia Policy period (1901–1973) still remain only 'sparsely covered' (Khoo and Noonan 2011: 92). Researchers have noted the need to include dynamic notions of movement, transnational networks and more in-depth community analyses into research on the period (Cushman 1984; Chan 1995, 2001; Reeves and Khoo 2011; and Reeves and Mountford 2011). I argue that one of the most obvious gaps in scholarship is the lack of understanding of Chinese Australian female lives and experiences in the White Australia Policy period.

A survey of research across a range of disciplines—namely history but also demography, geography, literary studies and cultural studies—highlights a striking limitation in understanding the diversity of Chinese Australian experience via the exclusion of women (Kamp 2013). As Bagnall (2006: 16) has argued, 'scant attention has been given to the lives of the, admittedly few, Chinese women who were part of the Australian Chinese community—as wives, mothers, daughters, servants and workers'. As such, 'when they are discussed, Chinese women and the White and other non-Chinese partners of Chinese men and their families are primarily seen as oddities and exceptions in a population of single men' (Bagnall 2006: 16). More recently, I have argued that little progress has been made in including women's experiences and voices in the research tradition (Kamp 2013).

The invisibility of Chinese Australian females' lives in scholarship is due to three primary reasons. Firstly, there has been an assumption that Chinese Australian women were absent during the White Australia Policy era as an outcome of Australian immigration restrictions and Chinese cultural values that restricted women's mobility. As a result, it has been understood that the twentieth-century Chinese Australian community comprised a 'bachelor society' of Chinese men (Bagnall 2006; and Kamp 2013). This has been an assumption perpetuated over the past five decades in which interest in Chinese Australian historical experience has flourished. For example, in one of the earliest scholarly publications on the Chinese Australian experience, Choi (1975) presented historical census data that indicated the presence of Chinese Australian females between 1901 and 1966 yet continually and paradoxically pointed to the 'lack' of such women. This is strikingly seen in Choi's (1975) discussion of the decline in the male Chinese population and small *increases* in the number of Australian-born Chinese females post-1901 (which saw the gradual balancing of the sex ratio). Immediately following the presentation of such data, Choi (1975) stated, 'This lack of females resulted firstly in a very slow growth of the Australian-born full-Chinese population and secondly in the prolonged predominance of aged males' (p. 47).

Overlooking Chinese Australian females on the basis of their relatively small population size has not been confined to early studies such as Choi's. Indeed, in one of the more recent and well-known studies examining Chinese communities in the White Australia Policy period, J. Fitzgerald's *Big White Lie* (2007), the author justified his chief focus on the experiences of Chinese Australian men by drawing upon the gendered demographics of Chinese settlement in Australia over the nineteenth and twentieth centuries. After providing a brief overview of the numbers of male and female Chinese Australians between 1901 and the 1930s and emphasising 'the absence of accompanying females' (despite overt references to the increasing female population), Fitzgerald explained: 'In light of the gendered demographics of Chinese settlement in Australia over the nineteenth and early twentieth centuries, the present study is concerned chiefly, albeit not exclusively, with Chinese Australian men' (Fitzgerald 2007: 14). Interestingly, the image on the front cover of the monograph is that of a Chinese Australian family—including female and male members.

The perpetuated assumption of female absence in academic scholarship is related to the second reason why Chinese Australian female lives have yet to be comprehensively examined. This reason is that broader migration and Chinese diaspora scholarship often assumes that (Chinese) migrants of this period were male or focuses only on male-dominated systems of migration and settlement (see Kamp 2021). Broader migration paradigms have, until fairly recently, leaned towards understanding political and economic reasons for male migration as uncovered by the 'push–pull' approach (Menjivar 2005; Pessar and Mahler 2003; and Ryan and Webster 2008). Within this research tradition, males have been conceptualised as primary actors in the migration process with women assumed to be dependants—either passively accompanying their independently mobile husband or following later as part of family reunification processes (Kershen 2008; Pessar and Mahler 2003; and Ryan and Webster 2008). The conceptualisation of the passive female migrant is also often grounded in the limited evidence and official documentation regarding their movements (Ryan 2003: 15). This manifestation of official patriarchy is itself a reflection of gender bias in the conceptualisation of the female migrant. Other feminist researchers have also argued that when the reality of women's economic migration has been acknowledged, it is often assumed 'that patterns of female migration are likely to mirror those of male migration, that is, that the potential differences between male and female migration are not likely to be of theoretical or empirical significance' (Thadani and Todaro 1979: 4, cited in Morokvašic 1984: 897). Thus, given the assumption that female mobility is defined by dependence on the male migrant, or is of no 'theoretical or empirical significance', female experiences and roles in the migration process have been traditionally overlooked (Pedraza 1991: 303). As Ryan (2003) argues, women have historically been 'invisible' in international migration stories, and academic scholarship has similarly overlooked their unique migrant experience in a case of 'gender blindness'.

Within the literature on historical Chinese migration to Australia, such male-dominated understandings have been perpetuated. In literature published prior to the 1980s, mentions of females were generally made in regards to legislative changes to the entry of wives and dependants (see, for example, Palfreeman's (1967) discussion of 'Families and dependants') and the resultant inability of male Chinese immigrants to have their wives settle with them in Australia. As Yong (1977) among others such as Yarwood (1964) and later Williams (1999: 52) have claimed, the legislative restrictions on female movement and entry into the nation resulted in an absence of women and subsequent lack of family life within Chinese Australian communities:

> Amidst the thriving of Chinese communities, there existed, however, a lack of family life chiefly resulting from the 1901 immigration restrictions which disallowed domiciled Chinese to bring their wives and families to build a permanent home in Australia. The lack of full family life may not appear to have undermined the structure of the Chinese communities in 1921, but it was to prove fatal in the 1930s and 1940s for the Chinese communities in New South Wales and Victoria.
>
> (Yong 1977: 221)

As I will demonstrate throughout this book via the examination of Chinese Australian females' presence and experience, Yong's claim of the fatality of Chinese communities by the mid-twentieth century was hugely exaggerated (and indeed incorrect). Subsequently, like many other early studies, it has helped establish a tradition of academic (and arguably public) erasure of Chinese Australian females and their families.

The gendered nature of the immigration and settlement process is made clear throughout the literature via terminology used, for example, 'wives of resident Chinese' (Yong 1977: 28) and anecdotes reported. The Poon Gooey case is illustrative of the latter. The use of this case to illustrate the inability of female Chinese (wives) to permanently settle in Australia reoccurs throughout the early literature (e.g., Yong 1977: 26 and Yarwood 1964: 80–81) and is often the only time a female is mentioned beyond demographic statistics. Central to this case was Ham Hop, wife of Poon Gooey, a highly respected greengrocer of Horsham, Victoria, who was allowed to stay and be with her husband in Australia between 1910 and 1913. Despite all efforts to extend Ham Hop's stay, she was eventually deported. This case has been used in the literature to emphasise and highlight the discrimination and absence of Chinese females as well as to 'illustrate the problems encountered in the administration of this compromise policy' (Yarwood 1964: 80). The actual presence of Ham Hop in Australia for those 2.5 years, her lived experiences, participation and contributions during that time are all overlooked—rather, focus has been on her deportation and absence.[4] Interestingly, while 'absent' women like Ham Hop (or Mrs Poon Gooey as she is usually referred to) were made somewhat visible in the early literature, very little has been mentioned of those women who managed to enter Australia and stay here as wives, daughters or independent migrants (or of those females who were born in Australia). Thus, female Chinese Australians do not figure in much of the literature as primary or independent migrants, or as pioneering or influential individuals. Rather, they have been constructed as dependants of men at the mercy of discriminatory legislation and government institutions.

In addition to Australian immigration policy, it is widely understood that the assumed absence of Chinese females in Australia was a reflection of economically driven migration patterns and patriarchal ideologies in China. Migration paradigms that centre on the role of male Chinese migrants as traders, coolies and sojourners, have focused on the economic motivations for male movement across the globe within the contexts of colonial exploits in South East Asia, the expansion of global markets and economic networks (see, for example, Wang 1991 and McKeown 1999). Ryan (2003) emphasised this gender imbalance in reference to Wang's (1991) explanation of 'modern' Chinese global migrations. She stated:

> Strictly speaking, the terms are 'gender neutral'. However, because the dis-course ignores women, and migrant patterns that are dominated by women are excluded in the dialogue, the terms are inscribed male by default.
>
> (Ryan 2003: 26)

Explanations of the gendered and economic nature of Chinese migration in the nineteenth and twentieth centuries, as purported by researchers such as Wang (1991) and McKeown (1999), are therefore characteristic of broader migration paradigms.

In the nineteenth- and twentieth-century context, the supposed absence of dependant Chinese wives in the 'Gold Mountain' diaspora countries of Aus-tralia, New Zealand, Canada and the United States has also been directly linked to women's position within the 'traditional' Confucian family system which denied them the opportunity to be independently mobile (see, for example, Choi 1975). This system is formed around gender and generational hierarchies in which men are superior to women and offspring are subordi-nate to their elders. Stemming from the theory of *yin* and *yang*—the two pri-mary principles of the universe—these inequalities are viewed as natural and pre-ordained. As Gates (1989) explains, individuals can outlive the elders to whom they are subordinate, can challenge political authorities by replacing them, but 'the subordination of women to men [is] seen as inevitable biolog-ical destiny' (p. 801–802). According to Confucian philosophy, males have familial authority and it is their duty to maintain family prosperity, marry and produce offspring. In contrast, the female role is defined by the 'three obedi-ences': obedience to the father, obedience to the husband after marriage and obedience to the eldest son after the husband's death. In this way, females should play a submissive role to their father, husband and eldest son, but more specifically, 'to defer to their husbands in decision-making, to put their hus-bands' career development ahead of their own needs, to be responsible for all the household chores, to nurture and care for their children and to look after elderly parents [in-law]' (Ryan 2003: 61).

The collectivist and patriarchal nature of the ideal Confucian system in the nineteenth- and twentieth-century context saw the continuation of husbands' family and male descent line being paramount. Thus, while Chinese men ven-tured to Australia, New Zealand, Canada or the United States in search of wealth and prosperity, it is understood that their wives (generically known as 'Gold Mountain women') were left behind and relegated to lives of 'live-widowhood' (Ip 1995, 2002; Yung 1998; and Ip and Liu 2008). As Choi explained:

As far as the Chinese were concerned, even when there was no restriction on Chinese female migration, Chinese wives had not often migrated with their husbands. A petition for the repeal of the 'Act to Make Provisions for Certain Migrants' in 1856, explained that Chinese wished to leave their wives and children to look after their aged parents, and that Chinese women were too weak physically to travel over long distances.

(Choi 1975: 40)

In Choi's (1975) view, the sojourner system of migration favoured by Chinese families was, in itself, gendered and indicated a disinclination of Chinese men to bring their wives to Australia and/or settle permanently with or without them. Yarwood (1964) made similar assertions. By utilising population statistics (more particularly the small number of females) for the first decade of the twentieth century, Yarwood pointed out 'the disinclination of the Chinese [...] to establish families in Australia' (1964: 78). In other geographical contexts, Ip (1990, 1995, 2002), Yung (1998) and Ling (1993, 2000) have uncovered that there were exceptions, with many Chinese wives following their husbands to New Zealand and North America. In Australia, however, such framings of the gendered nature of Chinese migrations, combined with immigration restrictions, have aided in the perpetuation of the 'absent wife' thesis.

The third reason why Chinese Australian women's lives have been underrepresented in research is methodologically rooted. As in most colonial nations, the Australian historical record is not only ethnocentric but also androcentric and as such, there is limited documentation of Chinese Australian women's lives. This has made female-centred (or simply female inclusive) investigations seemingly difficult (Kamp 2013, Bagnall and Martínez 2021b). Using the material available, researchers have focused on the development of policy, administrative changes, governmental pressures and attitudes of the dominant society towards Chinese Australians. Alternatively, they have focused on Chinese Australian men's social, economic and political experiences as documented in government records such as Census records, immigration records (including Certificates Exempting from the Dictation Test and related case files), naturalisation records and *Hansard* or non-government English-language records such as newspaper reports, police reports and anti-Chinese meeting minutes. The perspectives and experiences of Chinese Australians themselves have also begun to be uncovered from temple records, Chinese Australian newspapers, business records, Chinese consulate records, records from various Chinese Australian institutions and societies, as well as personal accounts in the form of letters and diaries, photographs, oral histories and interviews. While the experiences of men are at the centre of most of these records and have subsequently been the focus of most analyses, accounts of Chinese Australian female presence and experience can be found; we just need to refocus our gaze and examine the material with a keen interest in women's lives and experiences. As such, in *Intersectional Lives*, I draw upon some of the official records (primarily census data) and family photographs, but more stridently seek to understand the lived experiences of some of these women as remembered and told from their own perspectives. This emphasis on privileging the voices of research 'subjects' and making their lives visible in scholarly research is rooted in my postcolonial feminist approach to historical geography (which I detail in the following chapter). As such, it felt only right to begin this book with the voice of one of the women at the centre of my research.

Stella's account of growing up on Thursday Island is also pertinent as it touches on many issues that I will investigate in this book. Stella's account of her mother's life, grandmother's life and her own life not only highlights the enduring presence of Chinese Australian women during the White Australia Policy era but also the international and intra-national mobility of this group. It points to the existence of cultural pluralism decades prior to the official multicultural policies of the 1970s, the various roles and contributions of Chinese Australian women and how these differed over time, generations and in different spatial and social contexts. Stella's recollections also begin to suggest the interactions between Chinese Australian women and other members of the community (Chinese and non-Chinese) and lastly, illustrate the way in which the experiences of Chinese Australian women—both migrant and Australian born—were shaped by a range of constructed and intersectional identities (e.g., gendered, classed, racialised) that were negotiated within an amalgam of power structures across a variety of places and social contexts.

Relegating Chinese Australian females to the margins of academic research is problematic as it has not only perpetuated a 'gender-blind' history of Chinese migration and settlement in Australia. It is also problematic as it has perpetuated Orientalist assumptions of 'dependant' and 'oppressed' Chinese females who are victims of Confucian values regardless of historical, geographical or cultural context. In doing so, the social, economic and political participation of Chinese females in Australia has been overlooked and their contribution to Australian nation-building within the context of middle-class, White, male domination has been ignored. This book begins to correct such 'gender blindness' by weaving first-hand accounts of nineteen women's lived realities throughout the White Australia Policy era throughout the analyses. Their personal narratives, alongside family photographs and official records, are critically examined from a particularly postcolonial feminist perspective to reveal the ways in which Chinese Australian females experienced intersectionality and constructed geographies of cultural maintenance, economic participation, identity negotiation and belonging. Through such examinations, I argue that Chinese Australian females played important and specific roles in the development of Chinese Australian communities throughout the period and contributed to the broader development of the Australian nation—politically, culturally, economically and socially. In this way, the research I present in this book contributes to the uncovering of multiple and dynamic historical geographies of Chinese Australian women in the White Australia Policy era.

While this is the first monograph of its kind to solely focus on the lives of Chinese Australian females in this historical and geographical context, it must be noted that I am not the first or only person to be interested in Chinese Australian female lives in the late nineteenth and early twentieth centuries. There is a growing body of research that has begun to include Chinese Australian females in understandings of Australia's Chinese past. Much of this research positions female experience as an accompaniment to the primary analyses of male experience (see, for example, the work of Giese 1997; Macgregor 1998; Shen 2001; Tan 2003; and Wilton 2004). However, Loh's (1986) overview of the lives of ten Chinese women who migrated to Australia between 1863 and 1909, Couchman's (2004) summary of the lives of a select few women (Chinese and non-Chinese) of Melbourne's Chinatown during the federation period, Khoo and Noonan's (2011) examination of Chinese Australian women's contributions to wartime fundraising, Martínez's (2011) exploration of the political activities of members of the Darwin Kuo Min Tang in the early decades of the twentieth century (with a particular focus on the role of women in politics) have been pioneering. More recently, Gassin's (2021) insights of the debutante's of the Dragon Ball era and the collection of Chinese women's histories of mobility between China and Australia (edited by Bagnall and Martínez 2021a) provides important impetus to 'emphasize Chinese Australian women as historical actors in their own right' (Bagnall and Martínez 2021b: 2; see also Fong 2021). This body of research has been pivotal in inserting women's lives and perspectives within the Chinese Australian history research tradition and highlighting the important ways the patriarchal bias in research topics and representation can be corrected. This parallels work conducted in other White-settler nations such as the United States (e.g., Ling 1993, 2000) and New Zealand (e.g., Ip 1990, 1995, 2002), which have also begun to examine Chinese female lives and legacies in migration and settlement histories. *Intersectional Lives* has been encouraged by this emergent body of research and hopefully facilitates a more strident insertion of female Chinese Australian (and other female) lives in understanding Australia's social, cultural, economic and political development. Given the relative infancy of research into the lives of Chinese Australian females prior to Australia's age of formal multiculturalism, there is still much to uncover and examine from a variety of disciplinary perspectives and theoretical and methodological approaches. Therefore, in the final chapter of this book, I provide various directions for future research.

The postcolonial feminist approach that I have utilised in my research is central to this book's contribution (and is discussed in more detail in the following chapter). My research epistemology has drawn together ideas and methods advocated by postcolonial and feminist scholars across the social sciences, but particularly from the discipline of geography. This includes the privileging of female-centred research, research 'from below' and 'subaltern'[5] voices, as well as the acknowledgement of multiple, intersecting identifications and diversity of experience. Among other things, this approach has facilitated an exploration of female Chinese Australian's experiences and multiple and fluid identities within gendered, racialised and classed structures of power. It is via this postcolonial feminist perspective that I also ground the driving argument for this research— that the experiences of the many thousands of Chinese Australian women present in Australia between 1901 and 1973 are worthy of investigation and worthy of visibility in family, community and national historical geographies. While the Chinese Australian female presence was relatively small compared to their male counterparts, their stories and experiences of the politics of belonging and exclusion must be included in scholarship if we are to move towards a more inclusive understanding of Australia's past. Their lives and legacies must also be examined if we are to develop a deeper understanding of the challenges and debates facing contemporary multicultural Australia. By detailing my epistemological and methodological approach and considerations in Chapter 2, I hope to re-inspire researchers within and beyond the academy to revisit and scrutinise the historical archives and colonial tools of documentation, overcome limitations of discipline-specific research traditions, and seek out means to insert the voices and perspectives of groups and individuals previously hidden (if not erased) from historical and contemporary narratives. While *Intersectional Lives* presents historically and geographically specific research, I believe the approach can be applied across various international contexts.

In the following section, I provide an overview of Chinese settlement in Australia in colonial times and throughout the White Australia period.[6] Included in this overview is a brief outline of the immigration restrictions specifically focused on the entry of Chinese wives and a summary of the ways in which Chinese women were indeed able to enter the Australian nation despite the discriminatory immigration legislation. Detailed (although male-oriented) accounts of legislation and movements of Chinese to and from Australia have been provided in existing research. My overview does not replicate these accounts but is, instead, a concise summary that functions as an important contextual backdrop to the following chapters. I conclude this chapter by outlining the structure of this book.

Chinese migrations, settlement and restrictive legislation in the Australian context, 1800–1973

Chinese have been present in Australia for much of the country's European history. In fact, written records (most notably the nineteenth-century diary of the English navigator Matthew Flinders) suggest that Chinese contact with the Australian continent may have dated earlier than European settlement—being linked to trade relationships with the Indigenous populations of the continent's north (Fitzgerald 1997: 12). The relationship between Chinese traders and Australia continued in the fledgling colonies, with inhabitants of Sydney awaiting luxury imports from Canton as early as the turn of the nineteenth century (Fitzgerald 1997). While a few Chinese were included in the small number of free settlers and labourers that arrived in Australia at the beginning of the 1800s, Chinese became most visible in the colonies with the introduction of Chinese 'coolie' labour in the late 1840s. With the cessation of convict transportation, Chinese coolies, primarily from Amoy,[7] provided cheap labour for the developing colonies (Choi 1975; and Markus 1994). In 1854, estimates indicated that 2,400 Chinese coolies had been introduced into the colony of New South Wales (Choi 1975: 18). Between the 1850s and 1870s, a more significant wave of migration related to the Victorian and New South Wales gold rushes occurred (Huck 1968; Markus 1994; and Fitzgerald 1997). This was primarily made up of migrants from rural areas of Guangdong Province in southern China[8] (Choi 1975). In the 1850s, these Chinese migrants formed the second largest migrant group in Australia after the British (Elder 2005). Between 1853 and 1857, the Chinese migrant population in Victoria increased from 2,000 to more than 40,000, while in Queensland in the early 1870s there was one person of Chinese descent to every ten persons of European origin (London 1970: 8).

Throughout the second half of the nineteenth century, objections to the arrival of Chinese to the colonies increased, as did hostilities towards them. Such anti-Chinese sentiment on the goldfields culminated in a number of riots— namely the 1854 riot at Bendigo Field, Victoria, and the 1861 riots at Lambing Flats, New South Wales. Both riots led to the prompt passing of legislation in the respective colonies which dealt exclusively with the restriction of Chinese immigrants (London 1970; Markus 1994). South Australia also introduced legislation in 1857 restricting Chinese migration, while Queensland responded to its increasing Chinese population with similar legislation in 1876 (London 1970; and Elder 2005).[9] By the late 1870s, miners of Chinese origin had almost entirely withdrawn from the Australian goldfields. Despite ongoing racial hostilities and colonial restrictions on Chinese immigration, those Chinese who withdrew from the goldfields did not necessarily leave Australia. Those miners who did not return to their homelands made new lives—bolstering the Chinese workforce in the agricultural/pastoral industry or joining those who had established market gardens, small general and grocery stores, laundries, cafes, boarding houses and cabinet making businesses in mining towns and larger cities (Choi 1975; and Macgregor 1998). In this way, a diverse Chinese Australian population made up of individuals from various districts, villages and clans in China and of diverse economic classes, educational backgrounds and social statuses continued to participate in Australian life (Bagnall 2006: 120).

There is no denying that the nineteenth-century Chinese population in Australia was predominantly male. In 1861, only 11 of the 38,258 Chinese in Australia were female. Three decades later in 1891, this number had risen to 298 females but still only represented 1.6 per cent of the 35,821 total Chinese in Australia (Choi 1975). This stark gender imbalance has been attributed to a number of factors. Firstly, colonial schemes for coolie migration nominated men as the chosen labour force needed for development (Ryan 2003: 26). Secondly, as explained in the previous section, China's sojourner system of migration rooted in Confucian values and family systems also favoured men and discouraged female emigration (Choi 1975; Macgregor 1998; and McKeown 1999). And lastly, travelling costs and the Australian colonies' restrictive immigration legislation provided little incentive for Chinese female immigration and made the reunification of Chinese husbands and wives extremely difficult (Yarwood 1964; Choi 1975; Loh 1986; and Couchman 2004).

By the turn of the twentieth century, it was a strong conviction among Australian politicians and the public alike that the various colonies needed to be united. The fantasy of a 'settler colonialism' (Wolfe 1994), in which those of European, but more specifically British descent, could own, occupy and work the land, was threatened by the continued presence of the 'internal Indigenous

Other' and the increasing presence of the 'external non-White Other', including Chinese (Elder 2003). Underpinning this national imagining was the 'doctrine of racial superiority' that, linked to Social Darwinism, asserted that non-Whites were inferior to the White race in mind, body and morality (Markus 1994; Tavan 2005; and Jupp 2007). A vision of a 'White Australia' was therefore deemed the only way in which the British peoples, who were spread across the continent and differed in class and religion, could be united as 'equals' (Markus 1979; Grimshaw et al. 1994). Together, they could create a racially pure country and, more systematically and successfully, protect themselves against the contaminating capabilities of inferior non-White 'Others'. As Carty Salmon MP proclaimed to the House of Representatives in 1901 in his support for nationalised immigration restriction:

> [We] have here in Australia an admixture of the best and choicest of the white races; we have here the progeny of men of adventurous spirit and sturdy frame, who have shown themselves by their actions to be imbued with patriotism of the loftiest kind, and it is in order to preserve all these elements that we desire to shut out from Australia the inferior races.
>
> (Salmon cited in Kamp 2010: 420)

The imagined White nation envisioned by Salmon, and shared by many of his contemporaries, was '[a] fantasy of a clearly bounded and inviolable national space' (Elder 2003: 223; see also Jupp 2007 on the 'myth' of White Australia) in which the superior 'White race' could live and prosper safely in the midst of neighbouring 'teeming hordes' (Markus 1994: 112).

In 1901, 'survivalist anxiet[ies]' about Australia's geographical proximity to 'land hungry' Asia (Walker 2005: 71) culminated in the *Immigration Restriction Act 1901*—one of the first pieces of legislation to be passed by the newly federated nation and the cornerstone of the White Australia Policy (Yarwood 1964; London 1970: 9). Via the 'dictation test' the Act unified colonial attempts to exclude Chinese (and other 'coloured races' such as Japanese, Indians and Polynesians) from entering the nation and thus aimed to prevent an 'invasion' from the 'East' (Elder 2003). In the two decades following the passing of the Act, departures of Chinese immigrants exceeded arrivals almost every year so that by 1933, there were no more than 15,000 Chinese resident on the continent; down from almost 30,000 in 1901 (Choi 1975; and Fitzgerald 2007). In its original and later amended forms, the *Immigration Restriction Act* remained the fundamental means of fulfilling the immigration objectives of the White Australia Policy until its abandonment by the Whitlam Labor Government in the early 1970s (Palfreeman 1967; Brawley 1995; and Walker 2005). It must be noted here, however, that the objectives of the White Australia Policy, as operationalised by the *Immigration Restriction Act* among other articles of legislation, were never fully realised—the ongoing presence of non-European immigrants and Indigenous Australian populations throughout the period continued to challenge the 'fantasy' of White Australia (Elder 2003: 223).

In its initial form, the *Immigration Restriction Act 1901* did permit the entry of wives and children of migrants who were not prohibited. However, two years after its inception, the clause concerning the entry of wives and dependents was suspended and finally repealed in 1905. In addition, wives of Chinese nationals who were resident in Australia were no longer permitted to stay in the country (unless, of course, they were of European descent or born in Australia). Thus, despite the fact that Chinese families more willingly allowed females, particularly those who were married, to emigrate due to cultural and social changes in Republican China in the 1920s and 1930s, Australian immigration law continued to restrict female Chinese presence in the nation (Choi 1975; Loh 1986). These restrictions on the entry of Chinese wives have reinforced general assertions that Chinese Australian women were absent throughout the White Australia period.

There were, however, means by which Chinese females were able to enter Australia in the first half of the twentieth century. The wives and children of well-established merchants were permitted temporary entry (usually six months) with extensions often granted and temporary permits, in some cases, being converted into permanent ones (Bagnall 2013). Sponsored students were also allowed temporary entry under Chinese passports. Illegal dealings were another possible avenue for entry. Some cases have been noted of corrupt officials supplying false identities that allowed entry into Australia (Williams 1999), but the trade of identities facilitated by brokers connected to companies with Chinese and Australian branches may have been more common. It was through such connections that the purchase of Birth/Naturalisation Certificates or Certificates of Exemption from the Dictation Test (CEDTs) of deceased Australian-born or non-returning migrants was made possible (a phenomenon briefly noted by Macgregor 1998 and Williams 1999). It must also be remembered that females who had entered the nation prior to 1901 and were naturalised/domiciled or were married to naturalised husbands had been permitted to remain in Australia throughout the period. There were also the Australian-born females who were an outcome of these unions and Chinese-Anglo relationships and who were, by legal right, British subjects/Australian citizens.

During the White Australia Policy era, Australian-born Chinese and those Chinese who were able to enter the nation under Certificates of Domicile or other exemptions were not immune to the discriminatory precepts of the policy. Restrictions on movements in and out of the country, business ventures and ability to reunite with family members, denial of citizenship and naturalisation rights, and institutionalised discrimination in areas such as employment were all commonly experienced by Chinese Australians in the period (Palfreeman 1967; Markus 1994; Fitzgerald 1997; Elder 2005; Fitzgerald 2007; and Jupp 2007). In the wake of the atrocities of the Second World War, overtly discriminatory practices and policies based on 'scientific' racism were replaced with policies of assimilation based on new theories of social disadvantage (Markus 1994; Haebich 2002, 2008; and Tavan 2005). A somewhat new 'imagined community' (as defined by Anderson 1983) based on 'racial' inclusion was therefore constructed on the proviso of cultural conformity to an 'Australian way of life' (Haebich 2002, 2008). Thus, the pressure to abandon Chinese cultural practices, traditions and identities was a lived reality for many Chinese Australians in the White Australia period.

While the White Australia Policy was still formally in operation until 1973, the post-war abandonment of overtly racist policies and practices saw dramatic changes in the Australian governments' views of Chinese immigration, and, in particular, the entry of females. In 1950, the Colombo Plan was introduced and it allowed the large-scale entry of assisted female (and male) tertiary students from the Asian region—most of who were of ethnic Chinese origin (Williams 1999). From the late 1950s, immigration legislation pertaining to the entry and settlement of non-Europeans was also relaxed and had major implications on Chinese entries. For example, in reaction to the non-European war refugees and displaced persons who had arrived in Australia, provisions for some non-Europeans already in Australia to remain on humanitarian grounds were made. Additionally, 'distinguished and highly qualified non-Europeans [were] admitted for indefinite stay[s]' and 'the conditions for the admission of persons of mixed descent [were] clarified and eased' (London 1970: 26). There were also changes to naturalisation laws. Those non-Europeans (including Chinese) who had been resident in Australia for 15 years or more were now able to apply for naturalisation as were those persons who had been permitted to stay in Australia without periodical extensions of their permits. Additionally, Chinese (and other non-European) spouses of Australian citizens were able to apply for naturalisation (on the same basis as European spouses) and naturalised Chinese were able to bring their families to Australia (London 1970; and Ryan 2003). With such changes to Australia's immigration laws, the Chinese Australian population increased dramatically and the sex ratio quickly equalised by the 1960s.

Half a decade later, the 2016 Australian census recorded over 1.2 million individuals of Chinese ancestry—the fifth most common ancestry (4 per cent) after English (25 per cent), Australian (23 per cent), Irish (8 per cent) and Scottish (6 per cent) (Australian Bureau of Statistics (ABS), 2017). Over half a million individuals were born in China (excluding SARS and Taiwan), the fourth most common birthplace (2 per cent) after Australia (67 per cent), England (4 per cent) and New Zealand (2 per cent), and closely followed by India (2 per cent) and the Philippines (1 per cent) (ABS, 2017). Within the group of Australians with Chinese ancestry, more than half were females (652,168 compared to 561,736 males). Recent Chinese migration to Australia (since the fall of the White Australia Policy in the 1970s) has been vastly different to that of previous eras in terms of size of migrant flows, types of migrant patterns, birthplaces of migrants and reasons for migration—such that 'a new narrative of 'Chinese' in Australia [has] had to be etched onto the old, and re-created' (Ryan 2003: 8). Despite such vast differences, the continuous and perpetual presence and contributions of individuals of Chinese ancestry in Australia since the 1840s indicate an ongoing, although nuanced and complex, Chinese Australian history. The many thousands of today's Chinese Australians, whether male or female, first or fourth generation Australian and originating from mainland China, Hong Kong, Singapore, Indonesia or Fiji, live the legacy of those Chinese Australians—or *are* in fact those Chinese Australians—who struggled and survived the discriminatory legislation and public scrutiny of the colonial and White Australia era. At the same time, recent global engagements with China are changing the Chinese

Australian landscape and Chinese diaspora more broadly. Growing scrutiny of mainland Chinese (PRC) investment and influence across the globe, increasing geopolitical tensions between China and Western nations such as Australia and the United States, and the racialisation of the COVID-19 pandemic have placed contemporary Chinese diaspora communities back into the public spotlight (see, for example, Hamilton 2018). There is therefore a pressing need to provide further academic insights into the ways in which Chinese Australian identities, migration and settlement have been lived, perceived and discussed. Examining links between the past and present provides important means of considering and addressing such current debates.

Book outline

This book is focused on three critical themes/contributions. These are driven by a strong feminist and postcolonial agenda to improve the status of Chinese Australian females in understandings of Australia's past and in women's continued lives and legacies in contemporary Australia. The first is new ways to conceptualise Chinese females in the diaspora as gendered, classed, culturally varied and racialised individuals with multiple identifications, multiple subject positions and different forms of agency and mobility. The second, in line with ongoing calls to revise the patriarchal research tradition of Chinese Australian research (Loh 1986; Chan 1995; Bagnall 2006; and Kamp 2013), is a revision of patriarchal understandings of Chinese Australian history and broader understandings of overseas Chinese migrations and settlement experiences. The third is the demonstration of how historical geography, informed by postcolonial feminist approaches, can facilitate more nuanced understandings of past (and present) times and places that include women's diverse experiences and contributions at the domestic, local, national and international scale. These three foci or themes run through and link the remaining chapters in this book.

In the following chapter (Chapter 2), I detail my postcolonial feminist approach, with particular reference to recent moves in the discipline of geography (and historical geography more specifically) to decolonise the discipline and privilege women's voices and experiences. I reflect upon the use of colonialist tools (such as census records) for postcolonial purposes, my own positionality as an academic geographer and other subject positions such as being a female Australian of Chinese (albeit mixed) heritage. I also reflect on the ways in which a focus on place and space allows us to reconfigure women's geographies of belonging and exclusion via examinations of links between private and public spaces. In doing so, I present historical geography and its postcolonial and feminist developments as an effective means of revising the tradition of Chinese Australian historical research, understanding the complexity of Chinese Australian women's lives in the White Australia Policy era, and utilising such understandings to facilitate a clearer understanding of the contemporary Australian context. In Chapter 2, I also establish the book's central theoretical and methodological position, thus setting up the framework for the empirical analyses across Chapters 3–6. Within this discussion, I posit intersectionality as a way to understand the diverse experiences and politics of exclusion and belonging of Chinese Australian women in White Australia. I argue that intersectional analyses of women's experiences are crucial to redressing androcentric and ethnocentric research approaches that have dominated Western research traditions and are pivotal in ensuring that 'monolithic' representations of 'oppressed' and 'victimised' Confucian woman are not perpetuated.

In Chapter 3, I begin to counter the existing invisibility of Chinese Australian women in academic research by closely examining historical census data from 1911 to 1966. In doing so, I reveal the extent of female Chinese Australian presence in White Australia and argue that the Chinese Australian population cannot be solely defined as a population of men. Beyond the gender divide, I also indicate the ways in which the population of Chinese Australian females was comprised of a diverse mix of age groups, birthplace groups, migrant status/generation, etc. Census data on birthplace, geographical distribution and migration patterns will be presented and I argue that Chinese Australian women were not only present in the nation throughout the White Australia period, but they were internationally and intra-nationally mobile and geographically dispersed across the Australian continent.

As the lived experiences of Chinese Australian women's intersectionality lies at the heart of this book, pertinent first-hand narrative accounts of migration and settlement experiences accompany the discussion of the official record. These narratives are used to ensure that Chinese Australian women are not simply presented as numbers on a census table but as individuals whose diverse experiences of mobility were shaped by age, class, gender and other subject positions such as migrant/Australia-born status and position/role in the family. The varied ways in which such intersections of identity limited or facilitated women's mobility is examined within the broader political contexts of White Australia (with its corresponding race- and gender-based immigration restrictions and other discriminatory legislation) as well as Chinese social and cultural systems that were deeply rooted in Confucian ideology. I argue that both macro-political and social contexts impacted the global movement of Chinese women as well as more micro-level family systems and migration decisions. This provides an important backdrop to explore first-hand accounts of Chinese Australian women's diverse roles, contributions and interactions in White Australia in the subsequent chapters.

Chapter 4 moves away from national-level population/census statistics and begins a more strident focus on the lived realities of individual Chinese Australian women at the household scale. By drawing upon qualitative interview accounts of everyday life, this chapter draws attention to the home and family institution to examine the contributions, roles and interactions of Chinese Australian females in the domestic sphere. Framed by an understanding of the politics of the Confucian family system, the discriminatory mechanisms of the White Australia Policy, and their impacts on the intersectional identities of Chinese Australian females, this chapter firstly explores the influence of hierarchical family power structures on Chinese Australian females' childhood experiences such as access and achievement in education. This chapter then moves on to illustrate the diverse ways in which Chinese Australian mothers and adult daughters contributed to the family economy. An analysis of interview material indicates that some women dedicated their lives to unpaid work in the home, while other women worked in family businesses in subordinate positions. In some instances, Chinese Australian women took on more responsibility in these businesses in ways that challenged the 'front'/'back' gender divide. This economic participation reflected the need for Chinese Australian women to contribute to the survival of their families and experience roles and subjectivities that challenged the patriarchal division of gendered labour and space. The complex historical geography of Chinese Australian family systems in White Australia that is uncovered challenges traditional western feminist assertions that 'the home' is a universal site of female oppression and Orientalist assumptions that Confucian family systems were practised uniformly by all overseas Chinese.

Chapter 5 is dedicated to an examination of Chinese Australian women's experiences of the maintenance of 'traditional' Chinese culture as recounted and explained in their own words. By focusing on everyday cultural practices relating to language and food, and more marked traditions linked to annual festivities, this chapter indicates how elements of 'traditional' Chinese culture, beyond the Confucian family roles and responsibilities, were reproduced and adapted in unique ways. The chapter also highlights how these cultural practices were linked to a sense of 'Chineseness' in the lives of some Chinese Australian women. Through a postcolonial feminist analysis, this chapter positions homes as important spaces in which culture was maintained, and women, particularly mothers, as important and central players in the maintenance and reproduction of culture. While acknowledging obvious links between women's roles in the maintenance of culture and their gendered position within the Confucian family system, a postcolonial feminist perspective is used to argue that cultural maintenance was an act of empowerment and resistance within the context of assimilation. The information presented in this chapter, therefore, complicates and challenge understandings of the 'oppressed Chinese woman' in the Australian context.

Chapter 6 moves beyond the family and home to explore the interactions between Chinese Australian women and other members of the Australian community in more public spatial and social contexts such as schools/universities, workplaces and more general community spaces. In the first half of this chapter, recounted memories are drawn upon to examine the ways in which Chinese Australian women were constructed and viewed as racialised outsiders via their interactions with other Australians. I argue that cultural and/or racial identities—'Chineseness' or 'Australianness'—that were self-ascribed or imposed by others, fostered feelings of difference or similarity and were used as a mechanism to exclude or include women in various understandings of race, culture and nation. The ways in which gendered, classed and generational identities also marked women as 'outsiders' within their various social contexts are examined. In the second half of the chapter, the diversity of these interactive experiences is acknowledged by highlighting the friendships and social supports experienced by some Chinese Australian women. These experiences often fostered feelings of inclusion and belonging. From discussions with Chinese Australian women regarding their sense of Australianness, Chineseness and their interactions with their White Australian communities, interesting and complex issues regarding assimilation and, more particularly, the limits of assimilation, arose. These issues are addressed in the final part of the chapter.

The concluding chapter of the book summarises and synthesises the key findings of Chapters 2 to 6 and interrogates them in relation to the fundamental questions posed by the book. Why do Chinese Australian women's lives matter to the study of Chinese Australian experience and history? How does an understanding of Chinese Australian women's experiences in White Australia reconfigure our understanding of Australian national identity, nation-building and the contemporary politics of Australian multicultural identity? What new insights does intersectionality add to the study of Australian race relations and Chinese diaspora research? How do the methodological, conceptual and analytical insights provided by this case study contribute to broader feminist discussions regarding who speaks for whom and how androcentric and colonialist research approaches can be challenged for more inclusive research? This chapter also highlights the contributions of the presented research to historical geography, postcolonial feminist research, and Chinese Australian and diaspora studies. As this project sought to partially fill substantial gaps in the literature, the book concludes by suggesting various directions for future research.

Notes

1 Research participants' names are used throughout this book with consent. Where consent was not granted, pseudonyms have been applied.

2 I have not used the term 'White' or 'non-White' to perpetuate or advocate colonialist racialisation, but as a reflection of the racial discourses that informed the White Australia Policy and subsequent racial categorisation and understandings of national belonging in the historical context. Beyond these uses, I use the term 'people of colour' to denote racialised subjectivities beyond the position of Whiteness. I am acutely aware of the homogenising impact of these terms so where possible, I provide more specific cultural/ethnic attributions. There are inconsistencies within the Whiteness studies literature in regards to the capitalisation/non-capitalisation of the term 'White'. I have chosen to capitalise the term as it is used throughout this book as a proper noun denoting a socially constructed racial group not as a common noun denoting colour or shade (as also argued by Wachal 2000). Some authors I quote have used the lower case version of the word. In these instances, I have reproduced the term as written in the original text.

3 In line with my postcolonial feminist positioning, throughout the writing of this book, I have been wary of my use of the term 'female' to define individuals at the centre of my study. 'Female' is a term that denotes a gendered way of being or behaving (or 'performing' in the words of Butler 1990), rather than a category of biological sex (although it is often used in that way—see Eckert and McConnell-Ginet 2013). 'Woman' would perhaps be a better term of reference, however, 'woman' is also inscribed with connotations of gender. Furthermore, unlike the term 'female' which is relatively age-neutral, the term 'woman' is associated with adulthood; that is, one 'becomes' a woman after childhood/puberty. Because my research is focused on 'women' (as the book title suggests) of all ages—children and adults—I have used both terms, 'female' and 'woman', to identify particular age groups while acknowledging the complexity of such terminology. I use the term 'female' when speaking of the

entire age range (including children) or when distinctions between the age groups need to be made, for example, female children/adult females. I use the term 'women' interchangeably with adult females.

4 Bagnall (2021) has recently provided a ground breaking 'personal and intimate' history of Ham Hop.

5 For a definition of 'subaltern', see Harcourt and Escobar (2005: 2–3).

6 Some content in the following section has been previously published in Kamp (2021).

7 Amoy, also known as Xiamen, is a major city in the Fujian Province and lies on the southeast coast of the People's Republic of China. Captured by the British in 1841 during the first Opium War, Amoy was the first treaty port to be opened by the Treaty of Nanking (in 1842).

8 Guangdong is the southernmost province of China where most Chinese migrants travelling to Australia, New Zealand, the United States and Canada in the nineteenth and early twentieth century originated (Choi 1975: 3). More particularly, the areas of migrant origin in Guangdong included the Pearl River Delta region, Zhongshan, Dongguan, Gaoyao, Taishan, Kaiping, Zengchen, Nanhai and Panyu (Choi 1975: 35; Bagnall 2006: 119). Guangdong is often alternatively written as Kwangtung or known as Canton.

9 See, in particular, London (1970) and Choi (1975) for a discussion of the development of Anti-Chinese Leagues and uniform measures across the colonies to restrict the entry of Chinese.

References

Anderson, B. 1983, *Imagined Communities: Reflections on the Origin and Spread of Nationalism*, Verso, New York.

Australian Bureau of Statistics 2017, *2016 Census Quickstats*, Australian Bureau of Statistics, Canberra.

Bagnall, K. 2006, *Golden Shadows on a White Land: An Exploration of the Lives of White Women Who Partnered Chinese Men and their Children in Southern Australia, 1855–1915*, PhD thesis, University of Sydney.

Bagnall, K. 2011, 'Rewriting the history of Chinese families in nineteenth-century Australia', *Australian Historical Studies*, vol. 42, pp. 62–77.

Bagnall, K. 2013, *Finding Chinese Family Connections in the National Archives*, National Archives of Australia, Canberra.

Bagnall, K. 2021, 'Exeption or example? Ham Hop's challenge to White Australia', in K. Bagnall and J. T. Martínez (eds.) *Locating Chinese Women: Historical Mobility between China and Australia*, Hong Kong University Press, Hong Kong, pp. 129–150.

Bagnall, K. and Martínez, J. T. (eds.). 2021a, *Locating Chinese Women: Historical Mobility between China and Australia*, Hong Kong University Press, Hong Kong.

Bagnall, K. and Martínez, J. T. 2021b, 'Introduction: Chinese Australian women, migration, and mobility', in K. Bagnall and J. T. Martínez (eds.) *Locating Chinese Women: Historical Mobility between China and Australia*, Hong Kong University Press, Hong Kong, pp. 1–26.

Brawley, S. 1995, *The White Peril: Foreign Relations and Asian Immigration to Australasia and North America, 1919–78*, UNSW Press, Sydney.

Butler, J. 1990, *Gender Trouble: Feminism and the Subversion of Identity*, Routledge, London.

Chan, H. 1995, 'A decade of achievement and future directions in research on the history of the Chinese in Australia', in P. Macrgegor (ed.) *Histories of the Chinese in Australiasia and the South Pacific: Proceedings of an International Public Conference Held at the Museum of Chinese Australian History, Melbourne, 8–10 October, 1993*, The Chinese Museum, Melbourne, pp. 419–423.

Chan, H. 2001, 'Becoming Australasian but remaining Chinese: The future of the down under Chinese past', in H. Chan, A. Curthoys and N. Chiang (eds.) *The Overseas Chinese in Australasia: History, Settlement and Interactions*, Centre for the Study of the Chinese Southern Diaspora, Australian National University, Canberra, pp. 1–15.

Choi, C. 1975, *Chinese Migration and Settlement in Australia*, Sydney University Press, Sydney.

Couchman, S. 2004, 'Oh I would like to see Maggie Moore again: Selected women of Melbourne's Chinatown', in S. Couchman, J. Fitzgerald and P. Macgregor (eds.) *After the Rush: Regulation, Participation and Chinese Communities in Australia 1860–1940*, Otherland Press, Melbourne, pp. 171–190.

Couchman, S. 2011, 'Making the "last Chinaman": Photography and Chinese as a 'vanishing' people in Australia's rural local histories', *Australian Historical Studies*, vol. 42, pp. 78–91.

Cushman, J. W. 1984, 'A "colonial casualty": The Chinese community in Australian historiography', *Asian Studies Association of Australia Review*, vol. 7, no. 3, pp. 100–113.

Eckert, P. and McConnell-Ginet, S. 2013, *Language and Gender*, Cambridge University Press, Cambridge.

Elder, C. 2003, 'Invaders, illegals and aliens: Imagining exclusion in a white Australia', *Law text culture*, vol. 7, pp. 221–250.

Elder, C. 2005, 'Immigration history', in M. Lyons and P. Russell (eds.) *Australia's History: Themes and Debates*, UNSW Press, Sydney, pp. 98–115.

Fitzgerald, J. 2007, *Big White Lie: Chinese Australians in White Australia*, UNSW Press, Sydney.

Fitzgerald, S. 1997, *Red Tape, Gold Scissors: The Story of Sydney's Chinese*, State Library of NSW Press, Sydney.

Fong, N. 2021, 'The emergence of Chinese businesswomen in Darwin, 1910–1940', in K. Bagnall and J. T. Matrínez (eds.) *Locating Chinese Women: Historical Moblity between China and Australia*, Hong Kong University Press, Hong Kong, pp. 76–104.

Gassin, G. 2021, 'All eyes on you: Debutantes' explorations of Chinese Australian womanhood at the Dragon Festival Ball', *Australian Historical Studies*, vol. 52, no. 4, pp. 533–548.

Gates, H. 1989, 'The commoditization of Chinese women', *Signs*, vol. 14, no. 4, pp. 799–832.

Giese, D. 1997, *Astronauts, Lost Souls & Dragons: Voices of Today's Chinese Australians in Conversation with Diana Giese*, University of Queensland Press, St Lucia, Queensland.

Grimshaw, P., Lake, M., McGrath, A. and Quartly, M. 1994, *Creating a Nation*, McPhee Gribble, Ringwood Victoria.

Haebich, A. 2002, 'Imagining assimilation', *Australian Historical Studies*, vol. 33, pp. 61–70.

Haebich, A. 2008, *Spinning the Dream: Assimilation in Australia 1950–1970*, Fremantle Press, Fremantle, W.A.

Hamilton, C. 2018, *Silent invasion: China's influence in Australia*, Hardie Grant Books, Melbourne.

Harcourt, W and Escobar, A. 2005, *Women and the Politics of Place*, Kumarian Press, Bloomfield, CT.

Huck, A. 1968, *The Chinese in Australia*, Longmans, Croydon, VIC.

Ip, M. 1990, *Home Away from Home: Life Stories of Chinese Women in New Zealand*, New Women's Press, Auckland.

Ip, M. 1995, 'From gold mountain women to astronauts' wives: Challenges to New Zealand Chinese women', in P. Macgregor (ed.) *Histories of the Chinese in Australasia and the South Pacific*, Museum of Chinese Australian History, Melbourne, pp. 274–286.

Ip, M. 2002, 'Redefining Chinese female migration: From exclusion to transnationalism', in L. Fraser and K. Pickles (eds.) *Shifting Centres: Women and Migration in New Zealand History*, Otago University Press, Dunedin, pp. 149–165.

Ip, M. and Liu, L. 2008, 'Gendered factors of Chinese multi-locality migration: The New Zealand case', *Sites: A Journal of Social Anthropology and Cultural Studies*, vol. 5, pp. 31–56.

Jupp, J. 2007, *From White Australia to Woomera: The Story of Australian Immigration*, Cambridge University Press, Port Melbourne.

Kamp, A. 2010, 'Formative geographies of belonging in White Australia: Constructing the national self and other in parliamentary debate, 1901', *Geographical Research*, vol. 48, pp. 411–426.

Kamp, A. 2013, 'Chinese Australian women in White Australia: Utilising available sources to overcome the challenge of "Invisibility"', *Chinese Southern Diaspora Studies*, vol. 6, pp. 75–101.

Kamp, A. 2021, 'International migration and mobility experiences of Chinese Australian women in White Australia, 1901–1973', in K. Bagnall and J. Martínez (eds.) *Locating Chinese Women: Historical Mobility Between China and Australia*, HKU Press, Hong Kong, pp. 105–128.

Kershen, A. J. 2008, 'Preface', in L. Ryan and W. Webster (eds.) *Gendering Migration: Masculinity, Femininity and Ethnicity in Post-War Britain*, Ashgate, Hampshire, England, pp. xi–xii.

Khoo, T. and Noonan, R. 2011, 'Wartime fundraising by Chinese Australian communities', *Australian Historical Studies*, vol. 42, pp. 92–110.

Ling, H. 1993, 'Surviving on the gold mountain: A review of sources about Chinese American women', *The History Teacher*, vol. 26, pp. 459–470.

Ling, H. 2000, 'Family and marriage of late-nineteenth and early-twentieth century Chinese immigrant women', *Journal of American Ethnic History*, vol. 19, pp. 43–63.

Loh, M. 1986, 'Celebrating survival – An overview, 1856–1986', in Loh, M. and Ramsey, C. (eds.) *Survival and Celebration: An Insight into the Lives of Chinese Immigrant Women, European Women Married to Chinese and their Female Children in Australia from 1856–1986*, Published by the editors, Melbourne, pp. 1–10.

London, H. I. 1970, *Non-White Immigration and the 'White Australia' Policy*, Sydney University Press, Sydney.

Macgregor, P. 1998, 'Dreams of jade and gold: Chinese families in Australia's history', in A. Epstein (ed.) *Australian Family: Images and essays*, Scribe Publications, Melbourne, pp. 25–36.

Markus, A. 1979, *Fear and Hatred: Purifying Australia and California 1850–1901*, Hale and Iremonger, Sydney.

Markus, A. 1983, 'Chinese in Australian history', *Meanjin*, vol. 42, pp. 85–93.

Markus, A. 1994, *Australian Race Relations, 1788–1993*, Allen and Unwin, Sydney.

Martínez, J. 2011, 'Patriotic Chinese women: Followers of Sun Yat-Sen in Darwin, Australia', in L. Lee and H. Lee (eds.) *Sun Yat-Sen, Nanyang and the 1911 Revolution*, ISEAS Publishing, Singapore, pp. 200–218.

McKeown, A. 1999, 'Conceptualizing Chinese diasporas, 1842 to 1949', *The Journal of Asian Studies*, vol. 58, pp. 306–337.

Menjivar, C. 2005, 'Migration and Refugees', in P. Essed, D. T. Goldberg and A. Kobayashi (eds.) *A Companion to Gender Studies*, Blackwell, Oxford, pp. 307–318.

Morokvašic, M. 1984, 'Birds of passage are also women…', *International Migration Review*, vol. 18, no. 4, pp. 886–907.

Palfreeman, A. 1967, *The Administration of the White Australia Policy*, Melbourne University Press, Melbourne.

Pedraza, S. 1991, 'Women and migration: The social consequences of gender', *Annual Review of Sociology*, vol. 17, pp. 303–325.

Pessar, P. R. and Mahler, S. J. 2003, 'Transnational migration: Bringing gender in', *International Migration Review*, vol. 37, no. 3, pp. 812–846.

Reeves, K. and Khoo, T. 2011, 'Dragon tails: Re-interpreting Chinese Australian history', *Australian Historical Studies*, vol. 42, pp. 4–9.

Reeves, K. and Mountford, B. 2011, 'Sojourning and settling: Locating Chinese Australian history', *Australian Historical Studies*, vol. 42, pp. 111–125.

Ryan, J. 2003, *Chinese Women and the Global Village*, University of Queensland Press, St Lucia, Queensland.

Ryan, L. and Webster, W. (eds.) 2008, *Gendering Migration: Masculinity, Femininity and Ethnicity in Post-War Britain*, Ashgate, Hampshire, England, pp. 1–18.

Shen, Y. 2001, *Dragon Seed in the Antipodes: Chinese-Australian Autobiographies*, Melbourne University Press, Melbourne.

Tan, C. 2003, 'Living with 'difference': Growing up 'Chinese' in white Australia', *Journal of Australian Studies*, vol. 77, pp. 101–108, 195–197.

Tavan, G. 2005, *The Long, Slow Death of White Australia*, Scribe, Carlton North, Victoria.

Thadani, V. N. and Todaro, M. P. 1979, *Towards a Theory of Female Migration in Developing Countries: A Framework for Analysis*, Centre for Policy Studies Working Paper 47, Population Council, New York.

Wachal, R. S. 2000, 'The Capitalization of *Black* and *Native American*', *American Speech*, vol. 75, no. 4, pp. 364–365.

Walker, D. 2005, 'Australia's Asian futures', in M. Lyons and P. Russell (eds.) *Australia's History: Themes and Debates*, UNSW Press, Sydney, pp. 63–80.

Wang, G. 1991, *China and the Chinese Overseas*, Times Academic Press, Singapore.

Williams, M. 1999, *Chinese Settlement in NSW: A Thematic History/A Report For the NSW Heritage Office*, NSW Heritage Office, Parramatta.

Wilton, J. 2004, *Golden Threads: The Chinese in Regional New South Wales*, New England Regional Art Museum, Armidale in association with Powerhouse Publishers, Sydney.

Wolfe, P. 1994, 'Nation and miscegenation: The discursive continuity in the post-mabo era', *Social Analysis*, vol. 36, pp. 93–152.

Yarwood, A. 1964, *Asian Migration to Australia: The Background to Exclusion 1896–1923*, Melbourne University Press, Melbourne.

Yong, C. 1977, *The New Gold Mountain: The Chinese in Australia 1901–1921*, Raphael Arts, Adelaide.

Yung, J. 1998, 'Giving voice to Chinese American women', *Frontiers: A Journal of Women Studies*, vol. 19, pp. 130–156.

2 Intersectionality and postcolonial feminist geography as a way of inclusion

This book reconstructs past geographies (historical geographies) of Chinese Australian women and, as such, is situated in the 'borderland' of history and geography (to borrow a phrase from Darby 1953: 1). Within the sub-discipline of historical geography, feminist concerns and perspectives have begun to emerge, although this has been a relatively late and somewhat uneasy development (Morin and Berg 1999: 312). In 1988, Rose and Ogborn made the first call to historical geographers to consider feminism, claiming that 'the theoretical and empirical achievements of feminism in increasing our understanding of past societies have been almost completely ignored in the sub-discipline' (Rose and Ogborn 1988: 405). By ignoring feminist perspectives and approaches taken up in other areas of the discipline and across the social sciences and humanities, Rose and Ogborn (1988) argued that the gender-blindness and patriarchal assumptions within the sub-discipline would not be revised or corrected. Women would consequently remain marginalised and hidden from the geographies of the past: 'disappear[ing] from the reconstructed past as if they had never been' (Rose and Ogborn 1988: 405). According to Rose and Ogborn (1988), 'this is a political act [as it] demeans women's historical roles in society, the economy and the polity and so helps sustain their present oppression' (p: 405).

Other feminist geographers were quick to follow Rose and Ogborn's call. In North America, Kay (1989, 1990, 1991) critiqued the androcentrism of frontier historical geographies and called for a reassessment and insertion of women into historical narratives. Kay (1991) identified three dominant gender biases in existing North American historical geographies. They were: (1) the invisibility of women as subjects in historical geographies; (2) authors' androcentric and ethnocentric biases about people in the past; and (3) assumptions that societies and communities refer to both men and women (as well as White and coloured experiences; see also Kay 1990). According to Kay (1990, 1991), when women did appear in historical geographies they were not referred to as actors in the context of standard geographic themes relating to regional expansion and settlement. Instead, women were constructed within the context of sexual relations and reproductive roles as wives or families of men.

DOI: 10.4324/9781003131335-2

Through Kay's feminist critique, we can see some similarities between North American frontier historical geographies and existing literature examining Chinese settlement in Australia. The 'grand narrative' histories of Australia's White settlement have been broadened and stridently challenged to include non-European migrants and Indigenous Australians as well as more localised and individual histories. However, there is still much room to reassess our understandings of the White Australia period. The patriarchal assumptions that have permeated the regional historical geographies of North America have also dominated understandings of twentieth-century Chinese Australia. For example, like the women who participated in the regional development of the United States and Canada, Chinese Australian women have been largely rendered invisible in the understanding of Chinese migration and settlement in both urban and regional areas. This invisibility has been due to the construction of Chinese Australian history as a history of men and their economic activities, or because Chinese Australian men's activities and experiences have been generalised to include the experiences of women. When Chinese Australian women have been mentioned in the literature, they have been constructed as dependents of their male counterparts. Thus, following Kay's (1990) argument that this treatment of women 'leads to logical fallacies and factual contradictions' (p: 620; see also Kay 1991), I also argue (and will demonstrate in the following chapters) that existing literature regarding Chinese Australian history has largely perpetuated inaccurate understandings of Australia's past. Chinese Australian men's histories cannot be understood as complete histories of Chinese Australian migration and settlement. Like Kay (1990, 1991), I do not demean the importance of existing male-oriented literature in uncovering the experiences, roles, identities and contributions of Chinese Australian males to the settlement and development of Australia in the twentieth century. However, they must be identified and understood as such.

The feminist advances in historical geography established by Rose, Ogborn and Kay have inspired and informed the research presented in *Invisible Lives*. In this chapter, I will trace these advances to establish the central theoretical and methodological position of this book[1]. I will pay particular attention to the postcolonial feminist approaches which have moved to decolonise the sub-discipline (and geographical research more broadly), privileged women's voices and experiences, and utilised intersectionality as a conceptual and methodological tool. In this chapter, I also detail the way in which I have uncovered previously invisible lives and experiences of Chinese Australian women through interviews and re-reading of historical census data. This functions to demonstrate how and why postcolonial feminist approaches can be used in practical terms to bridge the gaps between history and geography, historical geography and postcolonial feminism. I hope the transparency of my approach and reflections inspire (and perhaps guide) other researchers to revisit the archives, draw on postcolonial perspectives, utilise intersectional approaches, and be creative in the ways in which information about seemingly obscure and hard to reach realms of our histories and geographies can be uncovered.

Revisiting the national epic and including 'home' and household geographies

In order to negate the androcentrism of North American historical geographies and near absence of interactions with feminist perspectives, Kay (1989, 1990, 1991) made a number of recommendations. She urged geographers to look towards historical research which had been influenced by feminism and were making important contributions to the understanding of women's roles and experiences in frontier expansion and development (Kay 1989). In her call to correct the gender imbalance, Kay (1991: 441) also advocated a reinterpretation of national epic style historical geographies. Historical geographers could, she argued, include women's roles in the specific themes of the national epic such as earning a living, working the land and migration. More attention could also be paid to the individual experiences of actors in the national epic rather than summarising male-centred expansion and settlement experiences. Additionally, rather than focusing on spaces and places at large scales which ensured a bias towards male public activities such as neighbourhoods and towns, Kay advocated a shift in attention towards smaller scales such as the household. This would allow women's (and others') roles in the domestic economy to be made visible. While being careful not to perpetuate definitions of the public/economic/ male and private/domestic/female spheres, Kay suggested a shift to smaller scales would allow an inclusion of women's economic activities. This dual economy model also provides space to consider the experiences of other ethnic groups who were excluded from the larger scale export economy (Kay 1991: 445).

Kay's call to pay attention to the more personal space of the house and home in historical geographies was followed by similar calls made by feminist geographers within and beyond the sub-discipline. As Blunt (2005) asserted, the domestic sphere is 'material and affective space' that is most often embodied by women and 'shaped by everyday practices, lived experiences, social relations, memories and emotions' (Blunt 2005: 506; see also Massey 2005; Blunt and Dowling 2006; Blunt and Rose 1994; Domosh 1998; and Blunt and Varley 2004). Importantly, postcolonial feminists have found that, like other 'places', the household is not fixed or neutral 'but a geographically and historically dynamic social institution in which gender is embedded and negotiated' (Chant 1998: 5). Given that post-colonial feminism focuses on women's multiple identities, different experiences and positions within various power structures and relations (i.e. intersectionality), postcolonial feminists in geography have thus explored the ways in which the 'home' and household are constructed differently for and by different women (Blunt and Rose 1994; Silvey 2006; see, for example, Yeoh and Huang 2000; Dwyer 2002; Chapman 2003; Blunt and Dowling 2006; Quinn 2010; and Ratnam 2018, 2020). This is based on the assertion that spaces and places 'are not neutral backdrops or uncomplicated stages for people's lives' or 'simply containers within which social relations develop' (Pratt and Hanson 1994: 25). Rather, '[p]laces are constructed through social processes and, so too, social relations are constructed in and through space' (Pratt and Hanson 1994: 25). Identities and experiences are thus constituted in different ways in different places (Pratt and

Hanson 1994: 6; see, for example, Radcliffe 1994; Peake 1993). Such understandings challenge dominant Western feminist understandings of the household as a universal site of patriarchal subordination.

Since Kay's initial prompts and alongside advances in postcolonial feminist geography, historical examinations of the complex ways in which the domestic sphere has functioned as a site of identity construction, survival and/or resistance have emerged. For example, Blunt's (2000) examination of British women's experiences in India during the Lucknow uprising/siege of 1957 highlighted the way in which 'home' functioned as a space of survival rather than 'embodied and domestic defilement' (p: 229). In a similar way, her following work explored the complex relationships between the concepts of home, identity and nationality for Anglo-Indians in colonial India. Blunt (2002) found that the domesticity of Anglo-Indian women (particularly mothers) often took on a political role—being central to a 'new' national Indian identity that served to resist British imperialism. Blunt has also examined the complex nature of experiences of 'home' among Anglo-Indian migrant women in the West (including Australia), which is particularly pertinent to this book. For example, by examining firsthand accounts of life in the late 1940s and 1950s Britain, Blunt (2008) argued that domestic challenges associated with settling into an unfamiliar culture and lifestyle were felt mostly by Anglo-Indian women, yet their adaptation to their new home can be viewed not simply as a narrative of domestic servitude but also as a story of survival and success. In light of these historical geographies of home and homemaking, the private realm can be viewed as a dynamic site of transformative potential (Blunt and Dowling 2006: 215)[2].

As these historical geographies highlight, feminist geographers have also been centrally concerned with bringing into focus the links between the public and private 'to challenge and reformulate the simple categorization of home with domestic and private spheres' (Blunt and Dowling 2006: 16). According to feminist geographical perspectives, no longer can the public and private spheres be imagined as disparate geographical locations in which the private is the site of the feminine, familial, domestic, and non-economic, completely outside and irrelevant to the public sphere of the masculine, work/production, and politics. Instead, critical examinations of the public/private dichotomy have found that the public and private are interdependent. Thus, home 'is best understood as a site of intersecting spheres, constituted through both public and private' (Blunt and Dowling 2006: 18; see, for example, England and Stiell 1997). As such, not only are 'the intersections of public and private in creating homes [...] geographically and historically specific', but are shaped by 'processes of commerce, imperialism and politics' (Blunt and Dowling 2006: 18–19). The experiences of Anglo-Indian women and British women in colonial India in the above-mentioned work of Blunt highlight the intrinsic links between the domestic and the political in geographies of the past. However, the historical blurring of public/private domains has also been examined by feminist geographers in regard to the important interconnections of home and work. For example, Cope (1998) has explored the relationship between home and work for wool mill workers in Lawrence, Massachusetts between 1920 and 1939 and the way in which gender and ethnicity functioned in the construction of place. Similarly, the work of McGurty (1998) on settlement house workers and their efforts at garbage reform in Chicago at the turn of the nineteenth century is also useful in highlighting the relationship between home and work in a specific historical and geographical context. These postcolonial feminist ways of conceptualising 'homes' and 'homemaking' are adopted in this book to provide a lens through which to assess the complex relationship between 'inside'/private and 'outside'/public spheres in Australia's past. For example, I will demonstrate in Chapter 4 that some Chinese Australian women physically blurred the boundaries between the public and the private by participating in family businesses, either in subordinate positions or in the 'front of house'. However, on a more conceptual level, when women's 'home-making' and unpaid work are acknowledged as important contributions to family economies/economic survival, the blurring of the public and private spheres and the empowering potential of the domestic realm can be seen.

Overcoming 'numerical paucity' and methodological challenges

Kay (1989) also argued that an inclusion of women in North American historical geographies could be achieved if 'historical geographers [do not] assume that low percentages of women in some regions correlated with obscurity' (p: 304). She illustrated this point by highlighting a small sample of women in frontier Montana who influenced public policies and had significant impacts on the much larger cohort of men in the region. Kay's (1989) findings raise an important and relevant point—the presence of a relatively small number of women does not warrant exclusion from research. As I discussed in the previous chapter, many researchers have justified their focus on Chinese Australian men (and disregard of Chinese Australian women) on the basis that there simply were not enough women present in Australia to warrant investigation. However, like the Montana women at the centre of Kay's study, the relatively small number of Chinese Australian women did influence the wider public, economic and political arena (see, for example, Couchman 2004; Khoo and Noonan 2011; Martínez 2011; Kamp 2018; and Fong 2021). And, beyond these 'large scale' impacts, I reiterate Loh's (1986) argument that the day-to-day activities and experiences of Chinese Australian women also need to be investigated for a more complete understanding of Chinese Australian historical geographies.

Despite Rose and Ogborn's (1988) general calls for more gender-balanced research, Kay's more specific critiques of North American scholarship, and the emergence of feminist historical geographies that examine the blurred boundaries between the public and private, historical geographies have largely continued to lack engagement with feminist perspectives and approaches. In Morin and Berg's (1999) terms, 'this subfield often seems like one of the last bastions of empirical geography complicit with masculinist language and values' (p: 315). Indeed, as is the case for feminist research across the social sciences and humanities, feminist historical geography remains on the margins of the sub-discipline. Domosh and Morin (2003) suggested that 'institutional problems' are not the only factor that has shaped the 'uneven travels of feminist historical geography' (p: 262). Methodological difficulties in incorporating historical analysis into feminist geographies are also an obstacle in bridging the two sub-fields. While both historical geography and feminist geography are largely based on qualitative research methods, the qualitative methods favoured by feminist geographers are ethnographic—'bottom up'—in nature. Thus, there is the obvious limitation for feminist geographers to conduct research 'from below' on long-deceased historical subjects. Domosh and Morin (2003) highlighted that this incongruity between research content and method therefore 'raises the thorny issue of for *whom* is historical geography research conducted?' (p: 262) and whether historical geography, in lacking specific subjects to emancipate, lacks political weight.

The methodological incongruities do not end there. While feminist geographers privilege ethnographic methods, historical geographies largely rely on historical documents housed in archives. These documents have generally been produced and stored by society's elites—White, heterosexual, literate men—and thus are limited in their ability to contribute to reconstructing historical geographies that include women and other previously (or continually) marginalised groups. As an alternative, non-traditional sources can sometimes be used to uncover women's historical geographies, but when that is not possible the challenge is to '[discover] appropriate strategies for approaching the archives and reading the silences embedded in them' (Domosh and Morin 2003: 262).

Moves to correct the gender lacuna

Notwithstanding such limitations, it would be imprudent to overlook the small and important body of feminist historical geographies that has begun to overcome such conceptual and methodological obstacles. In North America, this emergent body of research has been informed by postcolonial theories and has taken into account the interplay of multiple axes of difference (race, ethnicity, culture, class as well as gender) in past geographies (Morin and Berg 1999). For example, Schuurman (1998) examined the movements of First Nations women between their own communities and White settler society (in the form of co-habitation and marriage) in colonial British Columbia, Canada. By providing a postcolonial feminist reading of official colonial records, Schuurman's (1998) study can be seen as a "protest [...] against the [White, masculinist] narratives which have marked settler society" (p: 155). Not only did Schuurman insert First Nations women into understandings of colonial Canada which have previously, almost exclusively, focused on White men, but she also allowed their position as mobile actors in the colonial context to be revealed. In this way, Schuurman's study questioned the legitimacy of dominant (colonial) discourse and "unsettle[d] the history of colonial power (Schuurman 1998: 155). Other important feminist historical geographies in North America include the work of Gulley (1993), Kobayashi and Peake (1994), Morin (1995), Heffernan and Medlicot (2002), Dua (2007), and Zagumny and Pulsipher (2008).

Historical geographies in the United Kingdom and other White settler societies such as Australia and New Zealand have also followed the North American lead, engaging with feminist theories and methods to uncover particular geographies of migrant and Indigenous groups and females in national histories (Morin and Berg 1999; UK examples include Rose 1997; Blunt 2000; Tambou-kou 2000; McDowell 2004; and Wainwright 2007). It cannot be denied, how-ever, that feminist contributions to Australian historical geography continue to be marginal. Some examples of works in the small body of research include Teather's (1990) investigation of the use of literature and official documents to uncover the nature of working-class women's experiences in post-war inner-Sydney; Teather's (1992) examination of the role and impacts of the Country Women's Association of New South Wales between 1922 and 1992; Anderson's (1995) examination of the history of representational practices at the Ade-laide Zoo[3]; Gleeson's (2001) investigation of domestic space and disability in nineteenth-century Melbourne; and McKewon's (2003) historical geography of female prostitution in Perth. Within this literature, historical geographies that focus on or specifically include the experiences of non-White women in Australia are few and far between. Exemplary works include Fincher's (1997) discussion of immigrant women's representation in post-WWII Australian immi-gration selection and Ramsay's (2003) investigation of the complex negotiations of place identity in Charbourg's 'Chinatown' in the early twentieth century—in which Princy Carlo, an Indigenous woman of mixed Chinese descent, played a central role.

Invisible Lives contributes to this small, yet important, sub-disciplinary field and draws particular parallels to North American literature. For example, despite obvious differences between the Canadian study conducted by Schuurman (1998) and my own research on Chinese Australian women in White Australia, some commonalities are evident. Like Schuurman's (1998) investigation, my research focuses on a group of women who have been largely excluded from male-oriented understandings of the national development of a White settler society. Schuurman's (1998) study and my own both move beyond traditional assumptions of women's positions as dependants of men within the national nar-rative. Instead, women are positioned as active agents. While the First Nations women at the centre of Schuurman's research showed initiative and agency in their active pursuit of relationships with White men, my research investigates the active role Chinese Australian women played within the family and broader social/economic contexts. Both Schuurman (1998) and I also took into account intersectionality, that is, issues of patriarchy, race, class and gender, in examina-tions of the experiences of these previously invisible women.

Intersectional Lives focuses on a group of women whose historical lives are still within the reach of living memory. As such, it is uniquely positioned to demon-strate how some of the methodological barriers between historical geography and feminist geography can be overcome. It is also positioned to demonstrate how the perspectives and approaches of history and geography can be bridged for fruitful research (following on from the work of Clayton 2000; Anderson 2018; Gorman-Murray et al. 2018; Gibson and Warren 2018; and Darian-Smith and Nichols 2018). And finally, its methodological approach provides insights into the ways in which we can move away from colonising research traditions in history, geography and the broader social sciences.

In the remainder of this chapter, I detail what this approach looks like in real terms. I position intersectionality not just as a theoretical perspective, but as a methodological approach that can be used to avoid the Orientalist perils of 'monolithic Othering' and instead understand the diverse lived intersectional experiences of women. I also reflect on the use of colonialist tools (such as census records) for postcolonial purposes, the documentation of women's own voices and perspectives, and my own positionality. While this book as a whole indicates how these methods and approaches facilitate a nuanced understanding of the social and cultural construction of spaces, knowledges, identities and power relations in Australia's past, they can also be adapted and built upon for investigations of other national contexts and sites therein. I, therefore, hope to provide further impetus for researchers in other contexts to revise dominant understandings of identity and belonging across time and place, take on the methodological challenges of examining 'subaltern' lives of the past (and present), and consider the defining role of gender, race and class (among other subject positions) in our everyday experiences.

Postcolonial feminism and intersectional approaches in geography

Feminists, particularly anti-racist feminists and feminists of colour, have been increasingly engaged with postcolonial critiques established and developed by Edward Said, Homi Bhabha and Gayatri Spivak (e.g., Mohanty 1984, 2003; hooks 1981, 1989, 2000; Crenshaw 1989, 1991; Lake 1993; Kobayashi and Peake 1994; Raju 2002; Rajan and Park 2007; and Ahmed 2017). Emerging alongside postcolonial demands for constant interrogation and self-reflexivity, 'enabl[ing] a wholesale critique of Western structures of knowledge and power' (Mongia 1996: 2), these relatively recent feminist approaches no longer privilege women's shared gendered experience of patriarchal oppression. Rather, the focus is on women's multiple identities, different experiences, and marginalisations within various power structures and relations—what is now commonly referred to as 'intersectionality' (after Crenshaw 1989). Via intersectional frameworks and approaches (of which there are many, see Carbado et al. 2013), we are able to move beyond arguments of the centrality of gendered identity, no longer simplifying women as 'women first' and Black, working-class or lesbian second (Anthias and Yuval-Davis 1983; Johnson 2000; and Ahmed 2017). This shift is pertinently illustrated in hooks' discussion of American imperialism:

> Despite the predominance of patriarchal rule in American society, America was colonized on a racially imperialistic base and not on a sexually imperialistic base. No degree of patriarchal bonding between white male colonizers and Native American men overshadowed white racial imperialism. Racism took precedence over sexual alliances in both the white world's interaction with Native Americans and African Americans, just as racism overshadowed any bonding between black women and white women on the basis of sex. [...] In fact, white racial imperialism granted all white women, however

victimized by sexist oppression they might be, the right to assume the role of oppressor in relationship to black women and black men.

(hooks 1981: 122–123)

Within the context of American colonisation, shared experiences of gendered oppression did not unite White women and women of colour, and thus gender should not be privileged in analyses of these women's experiences. Rather, hooks suggested that the experiences of these women should be addressed in regards to their differing positions of power within the imperialist framework[4]. In this vein, feminists have moved towards the possibility of many 'feminisms' within the context 'of past and ongoing imperial power relations' (Johnson 2000: 5, 152).

Theories of intersectionality and developments in the full acknowledgement of power relations and differences between women (and men) have, once again, turned the feminist critique towards (Western) feminism itself. Postcolonial theorisations and critiques from feminists of colour within and beyond Western academic traditions have highlighted that inclusions of females of colour and females in other marginal positions in research have actually reinforced unequal power relations between White Western feminists and the females they seek to represent (see, for example, Mohanty 1984). This is because, in attempts to acknowledge difference, females of colour have been represented as a 'singular monolithic subject'—the 'third-world woman', or more recently, the woman of the 'Global South' (Mohanty 1984: 333; see also Gandhi 1998). There has been the continued assumption that the experiences of White, heterosexual, middle-class women are the 'norm' against which all other females are measured or compared and thus females of colour (and working class and homosexual, etc.) have been constructed as the 'Other'. In the case of the 'non-White female Other', constructions have particularly followed the lines of the 'exotic Other', the 'oppressed Other', or the 'victimised Other' (Anthias and Yuval-Davis 1983). Feminist geography has not been immune to such categorising assumptions. While feminist geographers have increasingly worked towards recognition of difference—spatial and social—between women/females, there have been criticisms that the White, western, heterosexual, middle-class assumptions of feminist geographers continue to focus research (Christopherson 1989; McDowell 1991; and Johnson 2000). In doing so, it has been argued that feminist geographers have continued to silence and/or marginalise females of colour and thus feminism is directly linked to imperialist processes (hooks 1981; Kobayashi 1994; and Johnson 2000: 6).

One way in which females' experiences can be assessed without resorting to imperialist representations of monolithic 'Otherness' is to understand and acknowledge female diversity in terms of social positions, roles, and differing/ intersecting marginalisations—the core of intersectional approaches. This can be achieved through the comparison of different communities or discussions of differences *within* communities in terms of class, religion, age and other social indicators (Ganguly 1995: 39). For example, the work of Mohammad (1999) on Muslim Pakistani women in Southern England highlighted the necessity of acknowledging and understanding differences within 'Other' groups. Her study found that experiences and views on education and employment, the upkeep of Pakistani traditions such as dress, female roles in the family and marriage, as well as willingness to accept or contest 'group' identity, were in no way homogenous. Rather, experiences and opinions differed according to age and whether the women were British-born and/or raised. Like the participants in Mohammad's (1999) study, the Chinese Australian women at the centre of this book differed in regards to a variety of social indicators (e.g., age, class, education/employment status, whether they are migrant or Australian-born, place of residence, marital status) and as such, their experiences differed. Therefore, informed by Mohammad's (1999) approach, Chinese Australian female experiences are examined in this book in relation to their varied identities and subject positions within multiple structures of power. In this way, I seek to acknowledge 'internal' differences and reduce the risk of 'Othering'.

Postcolonial and intersectional approaches have also prompted questions as to how (or if) researchers can ever unproblematically conduct research outside one's own class, racial and privileged position (Johnson 2000; see Peake 1993: 19–20; Kobayashi 1994, Raju 2002; and Staeheli and Nagar 2002). More particularly, the authority of White, middle-class women to represent 'those who remain on the margins' has been challenged (Kobayashi 1994; see, for example, hooks 1981, 1991; Lorde 1984; and Spivak 1988). Implicit in these concerns is the question of who is speaking for whom and the consequences for such representations. Researchers are no longer viewed as 'a disembodied, rational, sexually indifferent subject—a mind unlocated in space, time or constitutive interrelationships with others' (Grosz 1986). They occupy privileged positions in having the power to obtain information from the 'researched', to interpret that data, and disseminate it (Oakley 1981; and Winchester 1996). The research is more than likely a contributing factor to the researcher's academic credit and career. Thus, there exists a 'socio-political distance between the researcher and the "researched"' (Moss 1995: 82). Feminist and postcolonial academics have argued that when this 'gap' (as termed by Moss 1995) is exploited, discourses of colonisation and imperialist processes are maintained. This has been particularly highlighted when research 'subjects' are in more socially, economically, or politically disadvantaged positions than the researcher. But rather than abandon social research that focuses on women unlike themselves (ethnically, economically, sexually, etc.), White feminists (and other researchers engaged with the postcolonial critique and intersectional approaches) have come to understand the relationship between the researcher and the researched and how they can use their privileged positions to socially just ends (e.g., England 1994; and Kobayashi 1994, 2007).

In geography, feminist considerations of differences among women and ways of including 'Other' groups in research are particularly pertinent due to the discipline's historical ties to processes of colonisation. Just as mapping, 'discovery' of new places and people, and the documentation and description of such discoveries—precursors to the discipline—were historically (predominantly) the endeavours of men, such endeavours were also central to European imperialist expansion and colonisation. As Johnson (2000: 163) explains, map-making and 'the construction of 'othered' places through forms of description' facilitated acts of invasion and European settlement. Hudson (1977) has argued that modern geography itself continued this relationship with Empire. According to his essay, geography in the late nineteenth and early twentieth centuries was promoted 'to serve the interests of imperialism in its various aspects including territorial acquisition, economic exploitation, militarism and the practice of class and race domination' (Hudson 1977: 12 as cited in Driver 1992: 27; see also Wood 1992; Duncan and Sharp 1993; and Duncan 1993). Postcolonialism has therefore not only prompted feminist geographers to reinsert the silenced or 'subaltern' in research and consider the diversity of intersectional experiences, it has also prompted geographers to move towards revealing the ways in which the discipline has been grounded in acts of colonisation and oppression. Furthermore, the ways in which the discipline continues to be implicated in the process of colonisation—by ignoring or minimising issues of racism, ethnic difference and power structures in research—have been critically examined (Peake 1993; Smith 1994; and Johnson 2000: 162). For example, a growing body of postcolonial geographies has emerged in Australia and overseas that address ongoing power relations in and across place and space (Johnson 2000; see, for example, Keith and Pile 1993; Massey 1994; Radcliffe 1994; Sibley 1995; and Jacobs 1996). My own research project follows in a similar vein via its conceptual and methodological grounding in understandings of intersectionality. Of my particular concern is the way in which colonialist (and nationalist) discourses represent females, particularly females of colour.

There is an interesting tension to be noted here between locating, mapping and knowing as a colonialist move, and unearthing, hearing and bringing to the fore the subaltern experience as advocated by postcolonial geographers. This book plays on this tension by drawing attention to the ways in which 'colonialist' tools central to geography's tradition can be used for postcolonial agendas. That is, just as early geographic endeavours were concerned with the documentation of places and the people in those locations in order to bring such knowledge back to the Western world, this book documents the experiences of a 'subaltern' group of people in order to make them visible to the broader community. However, rather than utilising such documentation of people and places for the purpose of oppression and other colonialist motives, this book aims to give voice to the 'subaltern' group and provide a space in which they can be more carefully present in understandings of Australia's past. Thus, I point to the fine line within geographic research between tools of oppression and tools of emancipation, and highlight the powerful postcolonial capability of the discipline.

Listening to women's voices

One strategy called upon by feminists to ensure colonialist tendencies are not perpetuated is the provision of a space in which previously (or continually) silenced women (and men) can have their *own* voices heard (Radcliffe 1994; and Johnson 2000). This follows research 'from below' approaches used by historians and subaltern studies scholars who aim to allow previously silenced groups to be 'the subjects of their own history' (Chakrabarty 2005: 472; Sharp 2011). Geographers have also begun to engage with the experiences of subaltern groups such as migrants, however, it is feminist geographers who have particularly answered calls to insert the voices of the colonised (including 'subaltern women') into the discipline (Peake 1993; Johnson 2000; see, for example, Hopkins 2010; Quinn 2010; and McDowell et al. 2012).

The research at the centre of this book privileged this approach and utilised in-depth interviews with nineteen women who identify themselves as being 'Chinese' and who were resident in Australia prior to 1973 (whether Australian-born or migrant; see Appendix A). Participants were also invited to volunteer any personal documents such as family photographs, newspaper clippings or birth/marriage/migration certificates that would aid in the understanding of their experiences, many of which have been included throughout this book. Participants' years of birth ranged from approximately 1920 to 1952. As such, participants' ages at the time of interviews ranged from between 57 and approximately 80 years of age. In addition, six of the interview participants are foreign-born, with the remaining thirteen being Australian-born. Of the six migrant participants, three were born in Hong Kong, two in mainland China, and one in New Zealand. All migrant participants arrived in the post-war period—between 1947 and 1971. Australian-born participants included second and third generation Australians, with some having forebears (male and female) who migrated to Australia as early as the 1860s (see Appendix A). While this group of participants is relatively diverse, the limitations of this sample must be acknowledged. Women who migrated in the first half of the twentieth century are under-represented with no women who migrated in the pre-war period being represented, and only one participant having migrated prior to the establishment of the People's Republic of China in 1949. Information regarding the migrations of older-generation women was obtained in the form of Australian-born participants recalling the migration stories of their mothers or grandmothers. It is acknowledged that the passing of time may have impacted the 'accuracy' of participants' recollections and that experiences of the older generation of women (mothers) are recalled from daughters' perspectives (they are not first-hand accounts). Characteristic of qualitative research, the arrival at an 'objective truth' was not the aim of this project, but rather how individuals remember/perceive and voice their experiences.

Even with such limitations, interviews with these nineteen women provided an opportunity to record the voices and recollections of a group of individuals who had actually lived throughout the White Australia Policy period and gain insight into their feelings of identity and experiences across a variety of spatial and temporal contexts. This collection of first-hand accounts was not only a practical and efficient means of gathering experiential data which, in the words of Blunt and Dowling (2006) have 'remain[ed] hidden in more public historical narratives' (p: 34), but also provided important insight into their intersectional experiences and everyday realities throughout many stages of their life—in childhood at home and school, as young adults at university, as mothers and workers. By shifting focus away from the authoritative frameworks and views of the dominant 'White' society and instead allowing previously 'invisible' Chinese Australian women a space to speak for themselves, hegemony was acknowledged and anti-elitist approaches were privileged. As such, Chinese Australian women played an important role in the research process.

While my approach places great value on research 'from below' and more specifically, the interview method, I do acknowledge that other researchers (particularly advocates of participatory action research) would be critical of interviews as a means to generate substantive empower/power dispersal. They would point to the need for the 'researched' to have a say in the project aims, method, protocols, etc. (see Pain 2004), for research to fully '[affirm] people's right and ability to have a say in decisions which affect them and which claim to generate knowledge about them' (Reason and Bradbury 2006: 10 cited in Klocker 2008: 31). In addition, it must also be acknowledged that interviews, as a research method, have been particularly scrutinised for the unequal power relations that they can exploit. The 'myth of neutral detachment' (Kobayashi 1994) has been replaced by understandings that the interviewer and interviewee are both positioned subjects who enter into a social relationship shaped by broader societal power structures (Kearns 1991; and Smith 2006: 647). Stemming from broader understandings of the researcher-researched relationship, it is assumed that interviewers are more 'powerful' than their participants in ultimately having control of the interview process and, where the interviewer is also the author, the interpretation and dissemination of results (Pile 1991; Winchester 1996: 122; and Smith 2006: 248). Due to the potentially exploitative capabilities of the interview process, postcolonial and feminist researchers have pointed to the need for constant critical self-reflexivity (Dowling 2005). As McDowell (1992: 409) has argued, as researchers 'we must recognize and take account of our own position, as well as that of our research participants, and write this into our research practice rather than continue to hanker after some idealized equality between us'. By doing so, we can acknowledge and make visible the 'ways that knowledge is produced through the social relations of the interview' (Baxter and Eyles 1997: 510).

Such self-reflexivity in the research process has, however, indicated that the power relationship between interviewer and interviewee is not so rigidly defined. Scholars such as Smith (2006) and Pile (1991) have challenged unidirectional conceptions of power in interviews, claiming that 'the structures of power between the interviewer and the interviewed are complex and unstable' (Pile 1991: 464). As Smith (2006: 650) has argued in regards to interviews with societal 'elites':

> '[a]lthough, in terms of authorship, the researcher (where this is also the author) *does* exert significant levels of power in relation to the voices of the researched, this does not necessarily mean that the researcher is always in a position of power within the research encounter'.

My own experiences of the interview process confirmed these assertions and was a stark reminder of the complexity of the lived reality of intersectionality. Within the interview process, the information obtained in our interactions was generally unidirectional—I obtained information from the participant. Further-more, I occupied (and continue to occupy) a privileged position by having the power to interpret, organise and disseminate that information in the form of this book and associated publications, lectures and other mediums. However, given my epistemological and political convictions, the collaborative and non-exploitative potential of the interview method was maximised. It was made clear to the interview participants that I was advocating their presence, contributions and experiences in the nation throughout the White Australia period, and as such would use their voices and recollections as a means of inclusion. In this way, I maintained a sense that the participants and their recollections were piv-otal to my research project and broader public awareness and understanding. Furthermore, times and locations of interviews were chosen by participants, I maintained flexibility in the interview schedule to allow participants to speak about what they felt was important, and interview participants were able to edit their transcripts and provide additional information at later dates if desired. This ensured that informants were not exploited and maximised the postcolonial capabilities of the research.

The power relations between myself and participants were further compli-cated by our age, gender and ethnic backgrounds. Dowling (2005) has argued, as researchers,

> … [w]e have overlapping racial, socio-economic, gender, ethnic, and other characteristics. If we have multiple social qualities and roles, as do our informants, then there are many points of similarity and dissimilarity between ourselves and research participants.
>
> (p: 26)

This was a case in point during my own interview experience. As a female researcher who was in my mid-twenties at the time of interviewing and is of mixed Chinese-European ancestry, I was positioned as both insider and outsider—in a state of 'betweenness' to borrow the term from Nast (1994)—of the group of women that I was researching (see also Kobayashi 1994 and Dowling 2005). My role as researcher positioned me as an 'outsider' and my age and outward ('White') appearance further positioned me in this way (see also Kamp 2021b). It was perhaps because of my age and non-Chinese appearance that participants often took on a 'teacher' role, educating me about past times that I did not expe-rience and cultural aspects that I did not (or which they assumed I did not) know about. In these instances, I not only felt like an outsider 'looking in', but the interview participants took on an empowered position. The 'teacher'/'student' relation that was established between me and some of the interview participants was not only extremely useful in negating the exploitative potential of the inter-view method but also the collection of detailed and well explained qualitative data. My position as 'outsider' was, however, complicated by our shared gendered identity which, I am sure, contributed to the ease and comfort in which partic-ipants shared their experiences about gender roles and relationships, discrimi-nation, etc. In this way, I was positioned as an 'insider' with the ability to share some aspects of experience. My own Chinese ancestry brought further complexity to my insider/outsider status. Although I have Chinese ancestry, it is not clearly marked by my appearance. Therefore, I often found that participants were initially curious as to why I was researching Chinese women. It was usually in my responses to this question that I told participants of my Chinese ancestry. Interestingly, once women were knowledgeable of my Chinese heritage, I became, in a way, an insider—some participants asked me to share my own experiences and those of my ethnically Chinese mother or drew similarities between me and their children or grandchildren. As I aimed to maximise the collaborative and non-exploitative potential of the interview method, I always shared my own experiences when asked.

It is through such approaches that I assert that researchers can conduct postcolonial research outside their own subject positioning (arguably, we will never find another individual with identical subject positions as ourselves). It is clear that distance between White, Western, heterosexual, middle-class researchers and the 'others' whom they research have resulted in problematic divisions between researcher and researched. However, I believe it is important to emphasise that such distance has been the result of *constructed* (rather than innate) divisions between the 'self' (researcher) and 'other' (researched) (following Kobayashi's 1994 assertion). It is important for researchers to acknowledge their own positions and lived experiences of intersectionality; however, it is not beneficial to accept that differences between researchers and research participants prevent the conducting of fruitful research.

Re-examining historical census data

In addition to presenting the voices of Chinese Australian women themselves, in *Intersectional Lives*, I also re-examine census data—a source of information typically associated with research 'from above'. I utilise these official records as they provide the most accurate and efficient means of examining the diverse demographic characteristics of the total Chinese Australian female population throughout the twentieth century. Between 1901 and 1973, the formal years of the White Australia Policy period, eight national censuses were conducted by the Commonwealth (later Australian) Bureau of Statistics. The first was conducted in 1911 and the last of the period in 1971[5]. Included in the census records were information on age, education, occupation/employment, marriage status, geographical location, birthplace, and length of residence of racially defined 'Chinese' females[6]. Given the breadth of the national censuses (in terms of population coverage, time span and information collected), they can be used to piece together a broad national picture of female presence, experience and contributions within various social contexts across time (e.g., within families, schools, workplaces, and communities). Furthermore, given that many researchers have used demographic information obtained from censuses and other official records to discount the inclusion of Chinese Australian females in research and analysis—pointing to low numbers to justify their 'absence' claims—I deemed it essential to revisit official sources of information to provide statistical evidence of female Chinese Australian presence and thus correct the historical record. While nuanced or textured insights into the everyday lives of Chinese Australian women could not be obtained from these official records, their analysis did provide insight into the mobility, settlement, marriage, education and employment experiences of thousands of Chinese Australian women across the country. This information simply cannot be obtained elsewhere. Therefore, I was able to utilise the census records for postcolonial feminist purposes, that is, a means of 'putting women into' Australia's historical geography of Chinese settlement and national development.

Despite reservations about using racialised data, I decided that collecting and aggregating the data pertaining to females who were racially defined as 'Chinese' (both 'half-caste' and 'full-blood'), rather than those who were of Chinese nationality or birth, would be the most informative and relevant information. While the census of 1971 did ask respondents to identify their 'racial origin', the collected information was not provided in the subsequent census publications beyond Indigenous and non-Indigenous counts. This was perhaps a reflection of shifts in government attitudes towards racial identification. Given that I only examined census data that is contained in publicly available census reports, my analyses of racialised census data were therefore limited to the 1911–1966 date range (see Appendix B). I have also used this information as a marker of population size, and thus the presence of female Chinese Australians in the White Australia Policy era, for a variety of reasons. Firstly, while it is widely agreed that 'race' is a socially constructed category with no scientific or biological grounding, racial categories in censuses are important reflections of how social groups have been counted and classified in particular contexts. Just as Australian immigration policy can be understood as a way in which ideas of the White nation and those who belong in it have been constructed (see Fincher 1997), the racial categories in the Australian censuses reflect racial ideologies of the time and the way governments dealt with the 'colour issue'. Thus, the 'race' category provides insight into the ways in which Chinese Australian females were classified and racialised by authorities. Secondly, census data on 'race' have been directly linked to policy—in this case, perceived threats of 'coloured others' reflected in census data justified the White Australia Policy and its associated discriminatory legislation. In this way, the racial inventory would have had real impacts on the lives of Chinese women in Australia. Thirdly, despite the obvious inadequacy and racist underpinnings of the categorisations, 'race' provides the closest numerical reflection of those females who identified themselves as ethnically 'Chinese'. Using nationality or birthplace data would disregard those individuals who were Chinese nationals or China-born but did not define themselves as ethnically 'Chinese'. Similarly, utilising the latter categories as markers of 'Chinese' identity would overlook those 'Chinese' who were born in countries other than China or who were not Chinese nationals. Lastly, while categorisations of 'full-blood' and 'half-caste' are problematic on a variety of levels, their use in the censuses indicate the extent of racialisation in the period.

This use of data obtained from such colonialist classification systems may seem at odds with my postcolonial feminist epistemology. Earlier, I highlighted my awareness that geographical research has, in the past, utilised racist tools to classify people and places. I also noted that there exists an interesting tension between these colonialist tools and postcolonial ways of unearthing, hearing and bringing to the fore, 'subaltern' experiences. My use of racialised census data exemplifies this tension. I have actively sought to undertake research that is not complicit in processes of colonisation yet I have used racialised data to identify population numbers and other demographic characteristics of 'Chinese' females who were present in the nation during the White Australia period. This tension has been negotiated in a variety of ways. I have not used these data uncritically but acknowledged the constructed nature and power dynamics of racial classi-fication in the censuses. By doing so, I have not analysed census constructions/ definitions of race as individuals' own feelings of cultural or ethnic identity. Fur-thermore, classifications of 'half-caste' and 'full-blood' were used in the censuses to *exclude* undesirable 'Other' groups from the broader White community and subsequently identify them as divergent from the national identity. I, on the other hand, have used the definitions and classifications created by Australian statisticians as a means for the *inclusion* of a 'subaltern' group—hence my use of the identifying term 'Chinese Australian'. I have utilised the same data to illus-trate the diversity, presence and contributions of Chinese Australian women in the White Australia era. Therefore, I not only include this 'Other' group within the Chinese Australian community—commonly assumed to be a community of men—but also included them in broader understandings of national belonging.

My utilisation of racial data for *inclusive* research is particularly evident in my inclusion of 'half-caste' Chinese females. While some of these females may never have identified themselves as Chinese in day-to-day life, according to census defi-nitions they could never be considered part of the European/White population. Therefore, I, unlike Choi (1975), have included 'half-caste' Chinese females in my understandings of the total Chinese female population in Australia. Another reason for their inclusion is that they reflect an important component of Chinese presence in Australia, that is the formation of intimate relationships between Chinese and non-Chinese Australians and Chinese-European families who were integral to the development of Chinese Australian communities (see Bagnall 2011). Thus, the inclusive connotations of 'half-caste' have been utilised.

Other geographers have also taken up 'colonialist tools' for postcolonial pur-poses in their research. For example, Barnett (1998) conducted a postcolonial reading of nineteenth-century accounts of African exploration published by the Royal Geographical Society in order to draw attention to 'the historical processes that condemned certain knowledges, meanings and subjects to a place outside the field of what was considered intelligible, rational and disciplined scientific discourse' (Barnett 1998: 248–249). His postcolonial re-reading of geographical discourse can be viewed within the broader area of Indigenous geographies—an area in which moves to 'de-colonise' the discipline through the use of what were colonialist practices have become particularly strident. In their guest editorial for a special edition of *Geographical Research*, Johnson et al. (2007) also commented on the shifting nature of the discipline's engagement with Indigenous peoples. They asserted that '[w]hile defining [I]ndigenous peoples was once asserted as a clear and unambiguous process', a 'political imposition' (Johnson et al. 2007: 117), today the discipline's interactions with Indigenous people can and should be used to create postcolonial geographies which are 'concerned with breaking, and writing, the silences of the present as well as the past' (Gilmartin 2001: 35 quoted in Johnson et al. 2007: 119). The approach to census data I have used in this book parallels the postcolonial tenets of this emerging engagement with Indigenous geographies.

Historical geographies and contemporary contexts

It is often assumed that historical geographies are of no present relevance (Clayton 2000). However, by drawing connections between geographies of the past and their relations with the present, historical geographers have keenly argued the 'presentist' characteristics of historical geography research and thus advocated its relevance and utility (Clayton 2000). Morin and Berg (1999), for example, argued that:

> ...even geographies of the past are concerned with the present, even if they do not explicitly narrate a contemporary situation. Histories are almost always 'presentist'; they narrate the past in order to provide some understanding of the present.
>
> (p: 313)

In a similar way, Schein (2001) asserted that 'historical geography is poised to contribute a sense of the past in themes that pervade contemporary geographic thought' (p: 10). In making such connections between past and present it has been argued that historical geographers may be able to reinstate the sub-discipline as central to the study of geography more broadly (as it was considered between the 1930s and 1960s) and provide important contributions to the wider social sciences and humanities (Schein 2001).

This book follows such notions. Indeed, through the uncovering and analysis of the experiences of Chinese Australian women in the White Australia period, my research provides insight into past understandings of Australian identity, belonging and exclusion. However, this research not only fills gaps in understandings of past Australian geographies; it also facilitates a clearer understanding of the contemporary Australian context and contributes to current thematic concerns regarding identity and difference across space and place. The link between past and present is also particularly pertinent to my research project as the women at the centre have left legacies in contemporary Australia, be they cultural, economic/business, or familial. Indeed, many of the women themselves are still living and thus, in understanding their past experiences, we are able to more fully understand and address their and their descendants' contemporary situation. In this way, *Invisible Lives* ties in with much broader debates on national identity, diversity, cohesion, multiculturalism and Australia's place within the Asian region. It also ties in with work on the politics of recognition which assert that acknowledging and recognising previously silenced histories—such as Indigenous histories in Australia—can help processes of reconciliation (see, for example, Haebich 2000). While official apologies have been offered in other 'Gold Mountain' countries (Canada, the United States and New Zealand) to address previous discrimination and marginalisation of Chinese communities (see Li 2008; Beaglehole 2009; Blatz et al. 2009; and Edwards and Calhoun 2011), in Australia, such recognition of past injustices to this immigrant group remain overlooked (Han 2011; and Lowe Kelley 2011). In this way, the research I present in this book highlights the important role of feminist historical–geographical research in the Australian and broader global diaspora context.

Conclusion

The purpose of this chapter has been to establish the central theoretical and methodological position of *Intersectional Lives*. I endeavoured to do this by first grounding the research presented here within the context of postcolonial feminist critiques and advances in (historical) geography. More specifically, I detailed calls made by feminist geographers to include women's voices and experiences in research and illustrated ways in which conceptual and methodological challenges of working with 'subaltern' groups have been overcome. The research presented in *Intersectional Lives* has been inspired and informed by these previous critiques of 'gender blind' and colonial traditions of the discipline, and subsequent feminist advances. As such, this initial framing was essential in positioning this book's contribution and perhaps, as a historical geographer, I was intent on locating the book within its own historical geography of academic research. I then moved on to provide detailed insights into how I have responded to feminist critiques, detailing the conceptual and methodological approaches used. In doing so, I hope the transparency of my approach and reflections (as well as the outcomes of my research presented in the following chapters) are useful and encourage others to take up the call to engage with postcolonial feminist historical geographies.

Notes

1 This chapter is derived in part from Kamp 2018, 'Chinese Australian women's 'home-making' and contributions to the family economy in White Australia', *Australian Geographer*, vol. 49, no. 1, pp. 149–165, copyright the Geographical Society of NSW, available online: http://www.tandonline.com/DOI:10.1080/00049182.2017.1327783. This chapter is also derived in part from Kamp 2021a, 'International migration and mobility experiences of Chinese Australian women in White Australia, 1901–1973', in K. Bagnall and J. Martínez (eds.), *Locating Chinese Women: Historical Mobility between China and Australia*, HKU Press, Hong Kong, pp. 105–128 (Copyright Hong Kong University Press).

2 Beyond the sub-discipline of historical geography, see Burton's (2003) study of twentieth-century Indian women's memories of home in colonial India, and Johnson and Lloyd's (2004) examination of the relationship between Australian women's empowerment/disempowerment in the domestic realms of the 1940s and 1950s as constructed in women's magazines.

3 Anderson's contributions to the sub-discipline are not confined to the Australian context. See for example Anderson's (1991) examination of racial discourse in Vancouver's Chinatown.

4 hooks' theory of gendered and racial oppression in America was later adapted by the Australian historian, Jackie Huggins to describe the Australian colonial context. Huggins tailored hooks' words as follows: 'Australia was colonized on a racially imperialistic base and not on a sexually imperialistic base. No degree of patriarchal bonding between white male colonizers and Aboriginal men overshadowed white racial imperialism. White racial imperialism gave all white women the right to oppress Blacks—women and men' (Huggins 1987: 77).

5 The 'Federation Census' of 1901 was the last of the co-ordinated colonial censuses that were implemented in the later decades of the nineteenth century. While the date, the form, the questions and the occupation classifications were all standardised across the various state censuses of 1901, final results varied in their tabular presentation, for example, calculations of groupings. There were also subtle differences in who was included and excluded in the population (Wright 2011). It is for these reasons that I did not endeavour to include state census data for 1901 where it had not been previously aggregated.

6 A question regarding the 'race' of individuals was included in the Australian national census from the first national census in 1911 until 1966 and in a 'new guise' (to borrow the phrase from Horn 1987: 2) until 1981. For the first five national censuses (1911–1954), racial classification was dependant on self-reportage with non-European residents asked to classify their race according to the categories of 'full-blood' or 'half-caste', for example, 'full-blood' Chinese or 'half-caste' Chinese, as they had been in the colonial censuses up until 1901. Despite this self-reporting, the racial categories for data presentation were defined by the Commonwealth (later Australian) Bureau of Statistics.

References

Ahmed, S. 2017, *Living a Feminist Life*, Duke University Press, Durham.

Anderson, K. 1995, 'Culture and nature at the Adelaide Zoo: At the frontiers of 'human' geography', *Transactions of the Institute of British Geographers*, vol. 20, no. 3, pp. 275–294.

Anderson, K. 2018, 'Chinatown dis-oriented: Shifting standpoints in the age of China', *Australian Geographer*, vol. 49, no. 1, pp. 133–148.

Anderson, K. J. 1991, *Vancouver's Chinatown: Racial discourse in Canada, 1875–1980*, McGill-Queen's Press-MQUP, Montreal.

Anthias, F. and Yuval-Davis, N. 1983, 'Contextualizing feminism—gender, ethnic and class divisions', *Feminist Review*, vol. 15, pp. 62–75.

Bagnall, K. 2011, 'Rewriting the history of Chinese families in nineteenth-century Australia', *Australian Historical Studies*, vol. 42, pp. 62–77.

Barnett, C. 1998, 'Impure and worldly geography: The Africanist discourse of the Royal Geographical Society, 1831–73', *Transactions of the Institute of British Geographers*, vol. 23, no. 2, pp. 239–251.

Baxter, J. and Eyles, J. 1997, 'Evaluating qualitative research in social geography: Establishing 'rigour', in interview analysis', *Transactions of the Institute of British Geographers*, vol. 22, pp. 505–525.

Beaglehole, A. 2009, 'Looking back and glancing sideways: Refugee policy and multicultural nation-building in New Zealand', in K. Newman and W. Tavan (eds.) *Does History Matter?: Making and Debating Citizenship*, ANU E-Press, Canberra, pp. 105–123.

Blatz, C. W., Schumann, K. and Ross, M. 2009, 'Government apologies for historical injustices', *Political Psychology*, vol. 30, no. 2, pp. 219–241.

Blunt, A. 2000, 'Spatial stories under siege: British women writing from Lucknow in 1857', *Gender, Place & Culture*, vol. 7, no. 3, pp. 229–246.

Blunt, A. 2002, '"Land of our mothers": Home, identity, and nationality for Anglo-Indians in British India, 1919–1947', *History Workshop Journal*, vol. 54, no. 1, pp. 49–72.

Blunt, A. 2005, 'Cultural geography: Cultural geographies of home', *Progress in Human Geography*, vol. 29, no. 4, pp. 505–515.

Blunt, A. 2008, *Domicile and Diaspora: Anglo-Indian Women and the Spatial Politics of Home*, John Wiley & Sons, New Jersey.

Blunt, A. and Dowling, R. 2006, *Home*, Taylor & Francis, Abingdon, Oxon.

Blunt, A. and Rose, G. (eds.) 1994, *Writing Women and Space: Colonial and Postcolonial Geographies*, The Guilford Press, London.

Blunt, A. and Varley, A. 2004, 'Geographies of home', *Cultural Geographies*, vol. 11, no. 1, pp. 3–6.

Burton, A. M. 2003, *Dwelling in the Archive: Women Writing House, Home, and History in Late Colonial India*, Oxford University Press, Oxford.

Carbado, D. W., Crenshaw, K. W., Mays, V. M. and Tomlinson, B. 2013, 'Intersectionality: Mapping the movements of a theory', *Du Bois Review*, vol. 10, no. 2, pp. 303–312.

Chakrabarty, D. 2005, 'A small history of subaltern studies', in H. Schwartz and S. Ray (eds.) *A Companion to Postcolonial Studies*, Blackwell Publishing, Carlton, VIC, pp. 467–485.

Chant, S. 1998, 'Households, gender, and rural-urban migration: Reflections on linkages and considerations for policy', *Environment and Urbanization*, vol. 10, no. 1, pp. 5–21.

Chapman, T. 2003, *Gender and Domestic Life: Changing Practices in Families and Households*, Palgrave Macmillan, Basingstoke, Hampshire.

Choi, C. 1975, *Chinese Migration and Settlement in Australia*, Sydney University Press, Sydney.

Christopherson, S. 1989, 'On being outside "the project"', *Antipode*, vol. 21, no. 2, pp. 83–89.

Clayton, D. 2000, 'Historical Geography', in R. J. Johnston, D. Gregory, G. Pratt and M. Watts (eds.) *The Dictionary of Human Geography*, 4th edition, Blackwell, Oxford, pp. 337–341.

Cope, M. 1998, 'Home-work links, labor markets, and the construction of place in Lawrence, Massachusetts, 1929–1939', *The Professional Geographer*, vol. 50, no. 1, pp. 126–140.

Couchman, S. 2004, 'Oh I would like to see Maggie Moore again: Selected women of Melbourne's Chinatown', in S. Couchman, J. Fitzgerald and P. Macgregor (eds.) *After the Rush: Regulation, Participation and Chinese Communities in Australia 1860–1940*, Otherland Press, Melbourne, pp. 171–190.

Crenshaw, K. 1989, 'Demarginalizing the intersection of race and sex: A black feminist critique of antidiscrimination doctrine', *University of Chicago Legal Forum*, pp. 139–168.

Crenshaw, K. 1991, 'Mapping the margins: Intersectionality, identity, and violence against women of color', *Stanford Law Review*, vol. 43, no. 6, pp. 1241–1300.

Darby, H. C. 1953, 'On the relations of geography and history', *Transactions and Papers (Institute of British Geographers)*, vol. 19, pp. 1–11.

Darian-Smith, K. and Nichols, D. 2018, '"How our forebears lived": The modern nation, its folklore and 'living' heritage in twentieth-century Australia', *Australian Geographer*, vol. 49, no. 1, pp. 199–217.

Domosh, M. 1998, 'Geography and gender: Home, again?', *Progress in Human Geography*, vol. 22, no. 2, pp. 276–282.

Domosh, M. and Morin, K. M. 2003, 'Travels with feminist historical geography', *Gender, Place & Culture*, vol. 10, no. 3, pp. 257–264.

Dowling, R. 2005, 'Power, subjectivity, and ethics in qualitative research', in I. Hay (ed.) *Qualitative Reserch Methods in Human Geography*, Oxford University Press, South Melbourne, pp. 19–29.

Driver, F. 1992, 'Geography's empire: Histories of geographical knowledge', *Environment and Planning D: Society and Space*, vol. 10, no. 1, pp. 23–40.

Dua, E. 2007, 'Exclusion through inclusion: Female Asian migration in the making of Canada as a white settler nation', *Gender, Place & Culture*, vol. 14, no. 4, pp. 445–466.

Duncan, J. 1993, 'Sites of representation: Place, time and the discourse of the "other"', in J. Duncan and D. Ley (eds.) *Place/Culture/Representation*, Routledge, London, pp. 39–56.

Duncan, N. and Sharp, J. P. 1993, 'Confronting representation(s)', *Environment and Planning D*, vol. 11, pp. 473–486.

Dwyer, C. 2002. '"Where are you from?" young British muslim women and the making of home', in A. Blunt and C. McEwan (eds.) *Postcolonial Geographies*, Continuum International Publishing, London, pp. 184–199.

Edwards, J. A. and Calhoun, L. R. 2011, 'Redress for old wounds: Canadian Prime Minister Stephen Harper's apology for the Chinese head tax', *Chinese Journal of Communication*, vol. 4, no. 1, pp. 73–89.

England, K., 1994, 'Getting personal: Reflexivity, positionality, and feminist research', *Professional Geographer* vol. 46, pp. 80–89.

England, K. and Stiell, B. 1997, '"They think you're as stupid as your English is": Constructing foreign domestic workers in Toronto', *Environment and Planning A*, vol. 29, pp. 195–215.

Fincher, R. 1997, 'Gender, age, and ethnicity in immigration for an Australian nation', *Environment and Planning A*, vol. 29, pp. 217–236.

Fong, N. 2021, 'The emergence of Chinese businesswomen in Darwin, 1910–1940', in K. Bagnall and J. T. Matrínez (eds.) *Locating Chinese Women: Historical Moblity Between China and Australia*, Hong Kong University Press, Hong Kong, pp. 76–104.

Gandhi, L. 1998, *Postcolonial Theory a Critical Introduction*, Allen Unwin, Crows Nest, NSW.

Ganguly, I. 1995, 'Exploring the differences: Feminist theory in a multicultural society', *Hecate*, vol. 21, no. 1, pp. 37–52.

Gibson, C. and Warren, A. 2018, 'Unintentional path dependence: Australian guitar manufacturing, bunya pine and legacies of forestry decisions and resource stewardship', *Australian Geographer*, vol. 49, pp. 61–80.

Gilmartin, M. 2001, 'Making space for personal journeys', in P. J. Moss (ed.) *Feminist Geography in Practice: Research and Methods*, Blackwell, Oxford, pp. 31–42.

Gleeson, B. 2001, 'Domestic space and disability in nineteenth-century Melbourne, Australia', *Journal of Historical Geography*, vol. 27, no. 2, pp. 223–240.

Gorman-Murray, A., Kamp, A. and McKinnon, S. 2018, 'Guest editorial: Historical geographies down under', *Australian Geographer*, vol. 49, pp. 1–3.

Grosz, E. 1986, 'What is feminist theory?', in C. Pateman and E. Grosz (eds.) *Feminist Challenges: Social and Political Theory*, Northeastern University Press, Boston, MA, pp. 190–204.

Gulley, H. 1993, 'Women and the lost cause: Preserving a Confederate identity in the American deep south',

Haebich, A. 2000, *Broken Circles: Fragmenting Indigenous Families 1800–2000*, Fremantle Arts Centre Press, Fremantle, W.A.

Han, E. 2011, 'Chinese Australians call for an apology', *Sydney Morning Herald*, Sydney, www.smh.com.au

Heffernan, M. and Medlicot, C. 2002, 'A Feminine Atlas? Sacagewea, the suffragettes and the commemorative landscape in the American West, 1904–1910', *Gender, Place & Culture*, vol. 9, no. 2, pp. 109–131.

hooks, b. 1981, *Ain't I a Woman: Black Women and Feminism*, South End Press, Boston, MA.

hooks, b. 1989, *Talking Back: Thinking Feminist, Thinking Black*, South End Press, Cambridge.

hooks, b. 1991, *Yearning: Race, Gender and Cultural Politics*, Turnaround, London.

hooks, b. 2000, *Feminist Theory: From Margin to Center*, Pluto Press, London.

Hopkins, G. 2010, 'A changing sense of Somaliness: Somali women in London and Toronto', *Gender, Place & Culture*, vol. 17, no. 4, pp. 519–538.

Horn, R. V. 1987, 'Ethnic origin in the Australian census', *Journal of Population Research*, vol. 4, pp. 1–12.

Hudson, B. 1977, 'The new geography and the new imperialism: 1870–1918', *Antipode*, vol. 9, no. 2, pp. 12–19.

Huggins, J. 1987, 'Black women and women's liberation', *Hecate*, vol. 13, pp. 5–23.

Jacobs, J. 1996, *Edge of Empire*, Routledge, London & New York.

Johnson, J. T., Cant, G., Howitt, R. and Peters, E. 2007, 'Creating anti-colonial geographies: Embracing Indigenous peoples' knowledges and rights', *Geographical Research*, vol. 45, no. 2, pp. 117–120.

Johnson, L. 2000, *Placebound: Australian Feminist Geographies*, Oxford University Press, Oxford.

Johnson, L. and J. Lloyd. 2004. *Sentenced to Everyday Life: Feminism and the Housewife*, Berg, Oxford.

Kamp, A. 2018, 'Chinese Australian women's 'homemaking' and contributions to the family economy in White Australia', *Australian Geographer*, vol. 49, no. 1, pp. 149–165.

Kamp, A. 2021a, 'International migration and mobility experiences of Chinese Australian women in White Australia, 1901–1973', in K. Bagnall and J. Martínez (eds.) *Locating Chinese Women: Historical Mobility between China and Australia*, HKU Press, Hong Kong, pp. 105–128.

Kamp, A. 2021b, 'White Mongrel', in S. Sisko and V. Hutton (eds.) *Cultural Competence in Counselling and Psychology*, Palgrave Macmillan, Cham, Switzerland, pp. 70–73.

Kay, J. 1989, 'Western women's history', *Journal of Historical Geography*, vol. 15, pp. 302–305.

Kay, J. 1990, 'The future of historical geography in the United States', *Annals of the Association of American Geographers*, vol. 80, no. 4, pp. 618–621.

Kay, J. 1991, 'Landscapes of women and men: Rethinking the regional historical geography of the United States and Canada,' *Journal of Historical Geography*, vol. 17, no. 4, pp. 435–452.

Kearns, R. 1991, 'Talking and listening: Avenues to geographical understanding', *New Zealand Journal of Geography*, vol. 92, pp. 2–3.

Keith, M. and Pile, S. (eds.) 1993, *Place and the Politics of Identity*, Routledge, London.

Khoo, T. and Noonan, R. 2011, 'Wartime fundraising by Chinese Australian communities', *Australian Historical Studies*, vol. 42, pp. 92–110.

Klocker, N. 2008, *A participatory, action-oriented and youth-led investigation into child domestic work in Iringa, Tanzania*, unpublished PhD thesis, University of New South Wales, Sydney.

Kobayashi, A. 1994, 'Coloring the field: Gender, 'race,' and the politics of fieldwork', *The Professional Geographer*, vol. 46, pp. 73–80.

Kobayashi, A. 2007, 'Anti-racist feminism in geography: An agenda for social action', in L. Nelson and J. Saeger (eds.) *A Companion to Feminist Geography*, Blackwell Publishing, Oxford, pp. 32–40.

Kobayashi, A. and Peake, L. 1994, 'Unnatural discourse: "Race" and gender in geography', *Gender, Place & Culture*, vol. 1, no. 2, pp. 225–243.

Lake, M. 1993, 'Colonised and colonising: The White Australian feminist subject', *Women's History Review*, vol. 2, no. 3, pp. 377–386.

Li, P. 2008, 'Reconciling with history: The Chinese-Canadian head tax redress', *Journal of Chinese Overseas*, vol. 4, no. 1, pp. 127–140.

Loh, M. 1986, 'Celebrating survival - an overview, 1856–1986', in M. Loh and C. Ramsey (eds.) *Survival and Celebration: An Insight into the Lives of Chinese Immigrant Women, European Women Married to Chinese and their Female Children in Australia from 1856–1986*, Published by the editors, Melbourne, pp. 1–10.

Lorde, A. 1984, *Sister Outsider*, The Crossing Press, Trumansberg, New York.

Lowe Kelley, D. 2011, 'Chinese Australians owed apology for discrimination against forebears', *Sydney Morning Herald*, Sydney, www.smh.com.au

Martínez, J. T. 2011, 'Patriotic Chinese women: Followers of Sun Yat-sen in Darwin, Australia', in L. Hock Guan and L. Lai To (eds.) *Sun Yat-Sen, Nanyang and the 1911 Revolution*, ISEAS, Singapore. pp. 200–218.

Massey, D. 1994, *Space, Place and Gender*, Polity, Cambridge.

Massey, D. B. 2005, *For Space*, Thousand Oaks, London.

McDowell, L. 1991, 'The baby and the bath water: Diversity, deconstruction and feminist theory in geography', *Geoforum*, vol. 22, no. 2, pp. 123–133.

McDowell, L. 1992, 'Doing gender: Feminism, feminists and research methods in human geography', *Transactions of the Institute of British Geographers*, vol. 17, pp. 399–416.

McDowell, L. 2004, 'Cultural memory, gender and age: Young Latvian women's narrative memories if war-time Europe, 1944–47', *Journal of Historical Geography*, vol. 30, pp. 701–728.

McDowell, L., Anitha, S., Pearson, R. 2012, 'Striking similarities: Representing South Asian women's industrial action in Britain', *Gender, Place & Culture*, vol. 19, no. 2, pp. 133–152.

McGurty, E. 1998. 'Trashy women: Gender and the politics of garbage in Chicago', *Historical Geography*, vol. 26, pp. 27–43.

McKewon, E. 2003, 'The historical geography of prostitution in Perth, Western Australia', *Australian Geographer*, vol. 34, no. 3, pp. 297–310.

Mohammad, R. 1999, 'Marginalisation, Islamism and the production of the 'Other's' 'Other'', *Gender, Place and Culture*, vol. 6, no. 3, pp. 221–240.

Mohanty, C. T. 1984, 'Under Western eyes: Feminist scholarship and colonial discourses', *Boundary 2*, vol. 12, no. 3, pp. 333–358.

Mohanty, C. T. 2003, *Feminism Without Borders Decolonizing Theory, Practicing Solidarity*, Duke University Press, Durham, London.

Mongia, P. 1996, *Contemporary Postcolonial Theory: A Reader*, Arnold, London.

Morin, K. 1995, 'A "Female Columbus" in 1887 America: Marking new social territory', *Gender, Place & Culture*, vol. 2, no. 2, pp. 191–208.

Morin, K. M. and Berg, L. D. 1999, 'Emplacing current trends in feminist historical geography', *Gender, Place & Culture*, vol. 6, no. 4, pp. 311–330.

Moss, P. 1995, 'Reflections on the 'gap' as part of the politics of research design', *Antipode*, vol. 27, pp. 82–90.

Nast, H. J. 1994, 'Women in the field: Critical feminist methodologies and theoretical perspectives—opening remarks on 'women in the field', *The Professional Geographer*, vol. 46, pp. 54–66.

Oakley, A. 1981, 'Interviewing women: A contradiction in terms', in H. Roberts (ed.) *Doing Feminist Research*, Routledge and Kegan, London, pp. 30–61.

Pain, R. 2004, 'Social geography: Participatory research', *Progress in Human Geography*, vol. 28, no. 5, pp. 652–663.

Peake, L. 1993, '"Race" and sexuality: Challenging the patriarchal structuring of urban social space', *Environment and Planning D: Society and Space*, vol. 11, no. 4, pp. 415–432.

Pile, S. 1991, 'Practising interpretative geography', *Transactions of the Institute of British Geographers*, vol. 16, pp. 458–469.

Pratt, G. and Hanson, S. 1994, 'Geography and the construction of difference', *Gender, Place & Culture: A Journal of Feminist Geography*, vol. 1, no. 1, pp. 5–29.

Quinn, B. 2010, 'Care-givers, leisure and meanings of home: A case study of low income women in Dublin', *Gender, Place & Culture*, vol. 17, no. 6, pp. 759–774.

Radcliffe, S. A. 1994, '(Representing) post-colonial women: Authority, difference and feminisms', *Area*, vol. 26, no. 1, pp. 25–32.

Rajan, R. S. and Park, Y. 2007, 'Postcolonial feminism/postcolonialism and feminism', in H. Schwartz and S. Ray (eds.) *A Companion to Postcolonial Studies*, Blackwell Publishing, Oxford, pp. 53–71.

Raju, S. 2002, 'We are different, but can we talk?' *Gender, Place and Culture*, vol. 9, no. 2, pp. 173–177.

Ramsay, G. 2003, 'Cherbourg's Chinatown: Creating an identity of place on an Australian Aboriginal settlement', *Journal of Historical Geography*, vol. 29, no. 1, pp. 109–122.

Ratnam, C. 2018, 'Creating home: Intersections of memory and identity', *Geography Compass*, vol. 12, no. 4, pp. 1–11.

Ratnam, C. 2020, '(Re)creating home: The lived and gendered experiences of Tamil women in Sydney, Australia', in Kandasamy, N, Perera, N and Ratnam, C (eds.) *A Sense of Viidu: The (Re)creation of Home by the Sri Lankan Tamil Diaspora in Australia*, Springer, Singapore, pp. 119–139.

Reason, P. and Bradbury, H. (eds.) 2006, *Handbook of Action Research*, Sage Publications, London.

Rose, G. 1997, 'Engendering the slum: Photography in East London in the 1930s', *Gender, Place & Culture*, vol. 4, no. 3, pp. 277–300.

Rose, G. and Ogborn, M. 1988, 'Feminism and historical geography', *Journal of Historical Geography*, vol. 14, pp. 405–409.

Schein, R. H. 2001, 'Re-placing the past?', *Historical Geography*, vol. 29, pp. 7–13.

Schuurman, N. 1998, 'Contesting patriarchies: Nlha7pamux and Stl'atl'imx women and colonialism in nineteenth-century British Columbia', *Gender, Place & Culture*, vol. 5, no. 2, pp. 141–158.

Sharp, J. 2011, 'Subaltern geopolitics: Introduction', *Geoforum*, vol. 42, no. 3, pp. 271–273.

Sibley, D. 1995, *Geographies of Exclusion. Society and Difference in the West*, Routledge, London & New York.

Silvey, R. 2006, 'Geographies of gender and migration: Spatializing social difference', *The International Migration Review*, vol. 40, no. 1, pp. 64–81.

Smith, K. E. 2006, 'Problematising power relations in 'elite' interviews', *Geoforum*, vol. 37, pp. 643–653.

Smith, N. 1994, 'Geography, empire and social theory', *Progress in Human Geography*, vol. 18, no. 4, pp. 491–500.

Spivak, G. C. 1988, *In Other Worlds: Essays in Cultural Politics*, Routledge, London.

Staeheli, L. A. and Nagar, R. 2002, 'Feminists talking across worlds', *Gender, Place & Culture*, vol. 9, no. 2, pp. 167–172.

Tamboukou, M. 2000, 'Of other spaces: Women's colleges at the turn of the nineteenth century in the UK', *Gender, Place & Culture*, vol. 7, no. 3, pp. 247–263.

Teather, E. K. 1990, 'Early postwar Sydney: A comparison of its portrayal in fiction and in official documents', *Australian Geographical Studies*, vol. 28, no. 2, pp. 204–223.

Teather, E. K. 1992, 'The first rural women's network in New South Wales: Seventy years of the Country Women's Association', *Australian Geographer*, vol. 23, no. 2, pp. 164–176.

Wainwright, E. M. 2007, 'Constructing gendered workplace 'types': The weaver–mill-worker distinction in Dundee's jute industry, c. 1880–1910', *Gender, Place & Culture*, vol. 14, no. 4, pp. 467–482.

Winchester, H. P. M. 1996, 'Ethical issues in interviewing as a research method in human geography', *Australian Geographer*, vol. 27, pp. 117–131.

Wood, D. 1992, *The Power of Maps*, Guildford Press, New York.

Wright, B. 2011, *A History of the Australian Census of Population and Housing*, Australian Bureau of Statistics, Canberra.

Yeoh, B. S. A. and S. Huang. 2000. '"Home" and "Away": Foreign domestic workers and negotiations of diasporic identity in Singapore', *Women's Studies International Forum*, vol. 23, no. 4, pp. 413–429.

Zagumny, L. L. and Pulsipher, L. M. 2008, '"The races and conditions of men": Women in nineteenth-century geography school texts in the United States', *Gender, Place & Culture*, vol. 15, no. 4, pp. 411–429.

3 Presence, diversity and mobility

It cannot be denied that the *Immigration Restriction Act* implemented in 1901 and broader social exclusions associated with the White Australia Policy had significant impacts on the Chinese Australian population. Census statistics of Chinese in Australia for the first half of the twentieth century indicate that the population decreased from 29,627 'full' Chinese in 1901 to a mere 9,144 in 1947. However, in the post-World War II period, calls to abandon discriminatory legislation, moves towards more open immigration policy and 'unfavourable conditions'[1] in China saw the quick recovery of the Chinese population in Australia. In 1961, 20,382 'full' Chinese were documented in the census. As I discussed in Chapter 1, this fluctuating population has predominantly been characterised as a population of men. As Macgregor (1998) explained:

> Most of these men had wives in China. The wives either had had children before the husband left for Australia, or he would return every few years for a short while and more children would be conceived. In later years one or more adult sons would join the father. Sometimes this pattern of chain migration continued across several generations. The next generation would be born and raised in the village, but would live their working lives in Australia. Almost all returned to China to retire.
>
> (p: 26)

In this chapter, I will begin to present an alternative historical geography of Chinese Australian settlement by challenging the 'bachelor society' and 'absent women' theses[2]. I will do this by presenting official statistical evidence of the presence of Chinese Australian females throughout the period. A revised reading of the historical census data also uncovers important demographic characteristics of the female Chinese Australian population that highlight the diversity and mobility of this group. In line with my postcolonial feminist approach, throughout this chapter, women's own recollections of their migration and/or the migration histories of their female forebears in Australia are woven. These individual voices provide nuanced and important insights that cannot be obtained from census data. This includes how migration and settlement experiences were shaped by the broader political context of White Australia (with its corresponding race- and gender-based immigration restrictions and other discriminatory legislation), Chinese social–cultural systems (within and beyond the family),

DOI: 10.4324/9781003131335-3

and individuals' age, class, gender and other subject positions, and position/role in the family. By focusing on the three broad issues of presence, diversity and mobility, this chapter is the first of four that begins to make the lived intersectionality of Chinese Australian women visible in the White Australia period.

Evidence of female presence

Despite the decline in the broader Chinese Australian population between 1901 and 1947, a close inspection of census data indicates that the number of 'full' Chinese females steadily increased in this period. In 1901, females numbered 474 (1.6 per cent of the total 'full' Chinese population) compared to 29,153 males. By 1947, the female population had increased to 2,550 (27.9 per cent), and in 1961, the number of 'full' female Chinese in Australia had reached 6,145—almost one-third (30.1 per cent) of the 'full' Chinese population.

Including 'mixed' Chinese Australians in the total Chinese Australian population provides additional complexity to the picture of Chinese presence in Australia. Primarily, the number of Chinese Australians was larger across all census years when 'mixed' Chinese are included (see Table 3.1). What is most striking is the large increase in female presence within each census year. For example, in 1901, the female population total increases from 474 ('full' Chinese) to 2,008 (if both 'full' and 'mixed' are included). It is also important to note the relative gender balance in the 'mixed' Chinese population and the numerical consistency of this population across all census years. That is, while the 'full' Chinese population (male and female combined) experienced significant shifts in its size in the period, the 'mixed' Chinese population averaged around 3,000 individuals each year. Given that the majority of 'mixed' Chinese were Australian-born (see Table 3.2) and thus holders of Australian citizenship, this numerical consistency over time can be attributed to their relative immunity from discriminatory legislation.

Table 3.1 'Full Chinese' and 'Mixed Chinese' in Australia, 1901–1971[1]

Year	Full Chinese (%)			Mixed Chinese (%)			Total (%)		
	Male	Female	Total	Male	Female	Total	Male	Female	Total
1901	29,153	474	29,627	1,556	1,534	3,090	30,709	2,008	32,717
	(98.4)	(1.6)	(100)	(50.4)	(49.6)	(100)	(93.9)	(6.1)	(100)
1911	21,856	897	22,753	1,518	1,501	3,019	23,374	2,398	25,772
	(96.1)	(3.9)	(100)	(50.3)	(49.7)	(100)	(90.7)	(9.3)	(100)
1921	16,011	146	17,157	1,891	1,778	3,655	17,902	2,924	20,826
	(93.3)	(6.7)	(100)	(51.4)	(48.6)	(100)	(86.0)	(14.0	(100)
1933	9,311	135	10,846	1,901	1,602	3,503	11,212	3,137	14,349
	(85.8)	(14.2)	(100)	(54.3)	(45.7)	(100)	(78.1)	(21.9)	(100)
1947	6,594	2,550	9,144	1,599	1,351	2,950	8,193	3,901	12,094
	(72.1)	(27.9)	(100)	(54.2)	(45.8)	(100)	(67.7)	(32.3)	(100)
1954	9,150	3,728	12,878	1,404	1,276	2,680	10,554	5,004	15,558
	(71.1)	(28.9)	(100)	(52.4)	(47.6)	(100)	(67.8)	(32.2)	(100)

(Continued)

Table 3.1 (*Continued*). 'Full Chinese' and 'Mixed Chinese' in Australia, 1901–1971[1]

Year	Full Chinese (%)			Mixed Chinese (%)			Total (%)		
	Male	Female	Total	Male	Female	Total	Male	Female	Total
1961	14,237	6,145	20,382	1,648	1,538	3,186	15,885	7,683	23,568
	(69.9)	(30.1)	(100)	(51.7)	(48.3)	(100)	(67.4)	(32.6)	(100)
1966[2]	n/a	n/a	n/a	n/a	n/a	n/a	17,131	9,592	26,723
							(64.1)	(35.9)	(100)

1 For the years 1901–1954, figures for 'Full Chinese' have been appropriated from the category 'Full Blood Chinese' and 'Mixed Chinese' from the category 'Half-Caste Chinese' as defined by the Commonwealth (later Australian) Bureau of Statistics.
2 1966 census data are not available for the sub-categories 'Full Chinese' and 'Mixed Chinese'.

Source: Choi 1975; CBCS (1917, 1924, 1937, 1951, 1958, 1964, 1971).

Birthplace statistics

More information can be uncovered when we consider the birthplace of those Chinese Australian females who were present in Australia on census nights during the White Australia period. The majority of the 'mixed' Chinese female population were Australian born—99.1 per cent (n: 1,488) in 1921 and 90.4 per cent (n: 1,390) in 1961 (Table 3.2). This reflects the many unions between Chinese and White Australians in the colonial and White Australia period (as noted by Bagnall 2006, 2011). In the period before World War II, the majority of 'full' Chinese females were also Australian-born. For example, in 1911, 632 of the 892 (70.9 per cent) 'full' Chinese females were Australian-born. By 1933, this increased to 1,316 of 1,535 (85.7 per cent) 'full Chinese' females in Australia. Given that the number of foreign-born 'full' Chinese female population decreased in that same period (from 260 individuals in 1911 to 219 in 1933), we could speculate that the overall increase in the 'full Chinese' female population between 1911 and 1933 was primarily attributed to the birth of Australian-born females rather than immigration (as also argued by Inglis 1972 and Choi 1975).

Table 3.2 Australian-Born and Foreign-Born Chinese Females in Australia, 1911–1966

Year	Full Chinese			Mixed Chinese			Total[1]
	Australian-Born (%)	Foreign-Born (%)	Total Full Chinese (%)	Australian-Born (%)	Foreign-Born (%)	Total Mixed Chinese (%)	
1911	632	260	892	1,488	8	1,496	2,388
	(70.9)	(29.1)	(100)	(99.5)	(0.5)	(100)	
1921	906	237	1,143	1,755	16	1,771	2,914
	(79.3)	(20.7)	(100)	(99.1)	(0.9)	(100)	
1933	1,316	219	1,535	1,588	14	1,602	3,137
	(85.7)	(14.3)	(100)	(99.1)	(0.9)	(100)	
1947	1,804	746	2,550	1,296	55	1,351	3,901
	(70.7)	(29.3)	(100)	(95.9)	(4.1)	(100)	

(*Continued*)

Table 3.2 (Continued). Australian-Born and Foreign-Born Chinese Females in Australia, 1911–1966

Year	Full Chinese			Mixed Chinese			Total[1]
	Australian-Born (%)	Foreign-Born (%)	Total Full Chinese (%)	Australian-Born (%)	Foreign-Born (%)	Total Mixed Chinese (%)	
1954	2,222 (59.6)	1,506 (40.4)	3,728 (100)	1,199 (94)	77 (6)	1,276 (100)	5,004
1961	2,600 (42.3)	3,545 (57.6)	6,145 (100)	1,390 (90.4)	148 (9.6)	1,538 (100)	7,683
1966[2]	Australian-born	4,463 (46.5)		Foreign-born	5,129 (53.5)		9,592 (100)

1 The number of individuals that did not specify their place of birth in the censuses has not been included in this table. Discrepancies in 'Totals' therefore occur between Table 3.1 and Table 3.2.
2 1966 census data are not available for the sub-categories 'Full Chinese' and 'Mixed Chinese'.

Source: CBCS (1917, 1924, 1937, 1951, 1958, 1964, 1971).

The census statistics also indicate that between 1933 and 1947, the 'full' Chinese female population increased by 1,015 individuals: an increase of 527 'full' Chinese foreign-born females and 488 'full' Chinese Australian-born females (Table 3.2). Thus, for the first time since 1911, Australia not only experienced a net gain of foreign-born females, but a gain which marginally outnumbered the increase of the Australian-born cohort. In the post-war decades that followed, the impact of immigration on the female population was even greater. Between 1954 and 1961, the number of Australian-born 'full' Chinese females increased by 378. In comparison, the foreign-born 'full' Chinese population increased by 2,039. Unlike the pre-war years, we can assume from this information that the large increases in the 'full' Chinese female population from 1954 was primarily owing to the migration of foreign-born females. The presence, mobility *and* diversity of the female Chinese Australian population must therefore be acknowledged.

Census reports for the years between 1911 and 1961 documented countries of birth of female Chinese—both 'full' and 'mixed' Chinese. In these reports, at least 33 countries of birth besides Australia were presented (Table 3.3). These included countries in Australasia (such as New Zealand and New Guinea), Asia (such as China and Southeast Asian countries commonly associated with the Chinese diaspora, particularly 'Malaya', as well as India and 'Arabia'), the Americas (including the United States and British West Indies), Polynesia, Africa and Europe (including England, Norway, Italy and the USSR). The breadth of countries documented in the censuses indicates a diversity of migrant 'Chinese' females present throughout the period.

While we cannot know when using these data if birthplaces were indeed places of migrant departure, by tracking changes in birthplace over time, we can make tentative connections between the country of birth and the context of migration. For example, the position of Hong Kong as a British colony and 'the primary node through which Chinese overseas migration increased to unprecedented volumes' (McKeown 1999: 314) is most likely reflected in the consistent

Table 3.3 Country of Birth of Foreign-Born Chinese Australian Females, 1911–1961

Country of Birth	1911	1947	1954	1961
Australasia				
New Zealand	1	6	7	–
New Guinea	–	61	97	–
Papua	–	–	5	–
Unspecified	–	–	–	215
Europe				
England	2	1	3	–
Ireland	–	1	–	–
France	–	–	1	–
Germany	–	2	1	–
Norway	–	–	1	–
Italy	–	–	1	–
USSR	–	–	3	–
Unspecified	–	–	–	14
Asia				
Malaya	–	29	122	–
Hong Kong	12	95	246	–
Singapore	–	–	52	–
Straits settlements	2	26	–	–
Other British possessions	–	3	–	–
China	250	532	952	–
Japan	–	1	–	–
Timor	–	2	–	–
Philippines	–	1	2	–
Netherlands East Indies (Indonesia)	–	2	28	–
Arabia	–	3	–	–
India	–	–	1	–
Ceylon (Sri Lanka)	–	–	1	–
Other	–	3	21	–
Unspecified		–	–	3,359
Americas				
USA	1	1	5	–
British West Indies	–	1	2	–
Other	–	3	–	–
Unspecified	–	–	–	15
Africa				
Unspecified	–	–	–	4
Polynesia				
Fiji		2	13	–
Solomon Islands		11	–	–
New Caledonia		3	2	–
New Hebrides		2	4	–
Other		–	13	–
Unspecified		–	–	86
Unspecified	10	–	–	–
Total	278	801	1,583	3,693

Source: CBCS (1917, 1951, 1958, 1964).

presence of Hong Kong-born females between 1911 and 1954. The large number of China-born females counted in the 1954 census is perhaps associated with the mass emigration of mainland Chinese during the Chinese civil war and Communist takeover of China in 1949. While India, Sri Lanka, the Philippines and Indonesia were member countries of the Colombo Plan in 1954 (Malaysia and Singapore entering agreements in 1957 and 1966, respectively), it is difficult to make connections between the Colombo Plan and female immigration from 1954 birthplace statistics. It is even more difficult to make connections in 1961, as specific countries of birth were not specified. More generally, however, larger increases in migrant females in the 1960s and 1970s are most probably also related to the revised *Migration Act 1958* and Australia's more 'open door' policy.

Information regarding the birthplace of Chinese Australian women also provides important insight into the official status of national belonging. The status of the 'Australian citizen' was created through the *Nationality and Citizenship Act 1948*. Prior to the introduction of the *Act* in 1949, Australians could only hold the status of 'British subject' in accordance with the *Naturalisation Act 1903* and its later amended forms (Klapdor et al. 2009; and Couchman 2011). Whether of European or non-European descent, those born in Australia or elsewhere in the Commonwealth were, by default, officially classed as British Subjects. Under the *Naturalisation Act 1901*, 'aliens' could also be granted naturalisation by the Commonwealth and attain the rights and privileges of British subjects (Klapdor et al. 2009). However, in line with the rationale of the White Australia Policy, up until the 1950s, Australian residents from Asia, Africa or the Pacific Islands were prohibited from applying for naturalisation. These 'undesirable Others', including Chinese, were therefore denied the privileges of Australian-born subjects. In 1956, amendments to discriminatory policies allowed those previously undesirable immigrants to apply for naturalisation if they had been resident in Australia for 15 years (reduced to five years in 1966) (Jones 2005). Given the birthrights of Australian-born individuals, Australian-born Chinese were officially classed as British subjects/Australian citizens despite the discriminatory naturalisation policies. The number of Australian-born 'full' and 'mixed' Chinese females therefore indicate an official national 'belonging' that deviates from exclusionist ideologies that were inherent in the White Australia Policy.

Presence of wives and mothers

It cannot be argued that large discrepancies did not exist between the number of married male and female Chinese Australians in the first half of twentieth-century Australia. Choi (1975) explained that there was only one married 'full Chinese' female to every 26 married 'full Chinese' men in 1911. Without considering non-Chinese wives, Choi (1975) amongst others such as Macgregor (1998) suggested that the overwhelming majority of married 'full Chinese' men were separated from their wives who remained overseas. Within this context, it has been asserted that married Chinese females in Australia were 'scarce' and, given the overall low number of full Chinese females in Australia at the time, unmarried Chinese females were perhaps even fewer—forcing the continuation of overseas marriages between Chinese men who were residents of Australia.

Close inspection of the census record corrects these assumptions of absence or scarcity. For example, in 1911, there were 626 Chinese Australian wives. There were an additional 40 widows and one divorcee. In 1933, the number of married females had increased to 1,037 and there were an additional 203 widows and eight divorcees. By 1961, there were 2,425 married Chinese Australian women plus 70 who were married but permanently separated, 473 widows, and 40 divorcees. Although the population of single men outnumbered the cohort of Chinese Australian wives throughout the period (with single men numbering in the thousands in each census year), it is important to note that married women were present in Australia and, more particularly, in increasing numbers (see Table 3.4).

Information obtained through interviews with Chinese Australian women provides more nuanced evidence that Chinese Australian wives (and mothers) were present throughout the period. For example, two of the foreign-born interview participants (Lily and Daphne) entered Australia as wives of Chinese Australian men while several of the Australian-born participants recounted the histories of their mothers and grandmothers—most of whom were also migrant wives (which I will discuss further below). Thus despite strict immigration restrictions and cultural preferences to leave wives in ancestral villages, women's lived experiences and settlement histories highlight that migrant wives were not absent from the nation. In addition, of the 13 Australian-born interview participants, six married Chinese Australians (i.e. Ina, Stella, Doreen, Marina, Edith and Eileen). Mei-Lin, Marina, Edith and Kaylin also told me of their Australian-born female forebears who married Chinese Australians in the period. Focus within the literature on the gendered immigration context of the White Australia period has largely failed to acknowledge such Australian-born wives.

While Macgregor's (1998) study largely adhered to the 'absent wife' thesis, he did provide an important acknowledgement that marriage between Australian-born men and women of Chinese descent did occur. He explained:

> As Australian-born children of these [migrant] families grew to adulthood, their parents would seek brides and grooms on their behalf among other Chinese families in Australia. There were few Chinese in each town or city, so it was often hard to find a match locally. However, business and clan ties were strong within the total Chinese population around the colonies, so partners could be matched as far apart as Perth and Sydney, or Darwin and Launceston. As this practice was repeated over many generations, it is now common for descendants of many old Chinese-Australian families to be related to most of the other old families.
>
> (Macgregor 1998: 27)

This description of arranged marriages was not, however, reflected in the lives of the Australian-born interview participants of this study. Those who married Chinese Australians often met their husband at social functions—namely university or Chinese dances. For example, Susan and Ina met their husbands in this way and Doreen, who married in 1961 at the age of 20, recalled: 'My male cousin, who was a few years older than me, started taking me to Chinese dances and that's how I met my husband'. Other participants met their husband's through social

Table 3.4 Marriage Status of 'Full Chinese' and 'Mixed Chinese' Females in Australia, 1911–1961[1]

Marriage Status	1911			1933			1954			1961		
	Full Chinese	Mixed Chinese	Total	Full Chinese	Mixed Chinese	Total	Full Chinese	Mixed Chinese	Total	Full Chinese	Mixed Chinese	Total
Married	272	354	626	463	574	1,037	1,366	394	1,760	2,047	378	2,425
Never Married over 15	91[2]	412[2]	503[2]	410	482	892	1,009	234	1,243	2,289	224	2,513
Never Married under 15	465[2]	670[2]	1135[2]	563	424	987	1,048	486	1,534	1,392	770	2,162
Married but permanently separated[3]	–	–	–	–	–	–	27	25	52	38	32	70
Widowed	16	24	40	93	110	203	261	114	375	355	118	473
Divorced	0	1	1	0	8	8	12	20	32	24	16	40
Not Stated	53	40	93	6	4	10	5	3	8	0	0	0
TOTAL	897	1,501	2398	1,535	1,602	3,137	3,728	1,276	5,004	6,145	1,538	7,683

1 For the years 1901–1954, figures for 'Full Chinese' have been appropriated from the category 'Full Blood Chinese' and 'Mixed Chinese' from the category 'Half-Caste Chinese' as defined by the Commonwealth (later Australian) Bureau of Statistics.

2 The census records for 1911 did not indicate whether females who had never been married were over or under 15 years of age. Assuming that all females under 15 years of age were not of marriageable age and had therefore never been married, I have used information on age to extrapolate the number of females never married over and under the age of 15.

3 Category in 1954 and 1961 census only.

Source: CBCS (1917, 1937, 1958, 1964).

or family connections within the Chinese Australian community. For example, Eileen (who married at the age of 20 in 1966) explained: 'Who would think that many, many years later, when my eldest sister decided to come down to Sydney to work, she met up with her husband and because of that I met his cousin [my husband]?' Patricia, met her Australian-born husband through her church group:

> I was with the church group and this young fellow said come and play tennis with us. [...] So I went and played tennis and there I met my husband's older sister. [...] In those times we were all out looking for boyfriends and girl-friends. So she introduced me to her brother and at that age I thought oh, I better start looking around, too. [I] ended up going out with him.

These meetings and eventual marriages that took place in Sydney in the post-WWII period perhaps indicate that the arranged marriages described by Macgregor (1998) were less prominent in the second half of the twentieth century, particularly among the metropolitan population. It is unknown whether the marriages of participants' mothers/grandmothers in earlier decades followed the course of arranged marriages.

While the majority of interview participants were married to fellow Chinese Australians, women were not confined to endogamous relationships. Two interview participants—Sandy and Nancy—married non-Chinese Australians in the latter years of the White Australia period[3]. Sandy married her Anglo-Australian husband at the age of 26 in 1970. She explained that her parents, who never migrated to Australia but remained in Hong Kong, were initially unsupportive of the marriage:

> Oh, they didn't like the idea of marrying an Australian to start off with. So, when I rang them up and said I am getting engaged... 'Who? We don't know this boy! Forget it!' So in the end [my husband's] minister...he went to Hong Kong for a holiday or a conference or whatever. He went to talk to my parents—I didn't even know that—and came back. And so [they] changed their mind.

Nancy married her Irish Australian husband in 1966 at the age of 25 after they met in Canada. She explained her mother-in-law's reaction to the marriage in the following way:

> It was awful. He was an only child of Irish parents and I've read enough to know that Irish mothers in those days, they just...they were very possessive of their children especially, you know, he's an only child. So I think she was just shattered that he would marry a Chinese.

Thus, while her own parents were supportive of the marriage, the union was not approved by her mother-in-law who preferred the maintenance of endogamy.

Recent research uncovering the non-endogamous relationships in the Chinese Australian community have generally focused on the nineteenth- and early twentieth-century context and have understood that these relationships were between Chinese Australian men and non-Chinese Australian women (see Macgregor 1998; Williams 1999; and Bagnall 2006, 2011). This was probably largely due to the much larger number of Chinese Australian men in that period. However, Sandy and Nancy's marriages to non-Chinese men in the post-war period provide added complexity to the picture of Chinese Australian family life and, more particularly, Chinese Australian women's experiences as wives in the White Australia context.

Presence of unmarried Chinese Australian women

Census data pertaining to the 'conjugal condition' of Chinese Australian females also highlight the presence of unmarried women. For example, in 1911, there were 503 unmarried females of marriageable age—only marginally less than the number of married females. As the population of female Chinese Australians increased over the census years, so did the number of unmarried women. In 1933, the number of unmarried women of marriageable age increased to 892 (41.5 per cent) and in 1954 to 1,243 women (35.8 per cent). By 1961, there were just over 2,500 (45.5 per cent) unmarried females of marriageable age compared to 2,425 (43.9 per cent) married females (Table 3.4).

Three of the interview participants for this project—Kaylin, Mei-Lin and Margaret—can be included in these groups of 'single' women as all three never married during the White Australia Policy era, and in fact, remain unmarried. When I asked Margaret if there was any pressure put on her to marry during her young adulthood, she explained that pressure often came from her uncle:

> He wanted me to meet someone and I said, 'No'. And I said, 'Stop pressuring me, I'm not meeting anyone and that's it'. And I just kicked up such a stink that they didn't do it again, because I can kick up a stink.

Margaret later explained the reason why she did not want to marry was a matter of independence and prioritising her career, she recalled: 'I said, "Mum, look, I've got too much work to do. I've got too much things to do and I can't be tied down looking after someone, having a family. I can't do it"'. In contrast, Mei-Lin, remained in a de-facto relationship with her Anglo-Australian partner until his passing. They did not co-habit for the majority of their shared lives and never had children, but unlike Margaret, she felt no pressure from her family:

> …they [my parents] were very good about it. […] They were just happy that I was, you know, that I was happy […]. If they were disappointed they never let on. I don't think they were because it meant that I could come home more often and do things with them, that if I'd been married I wouldn't have.

Despite such differences in parental encouragement of marriage, both Margaret and Mei-Lin indicated that they made a conscious decision not to marry. On the other hand, Kaylin, who continued to live with family, made reference to external factors that limited her ability to meet potential partners. When I asked her if there was a reason why she never married, she responded by saying:

> I just think it was because we lived such an isolated sort of life, and it was, as I say, it was a serious sort of life. You just went out to go to school or do studies and come home and you'd…you were involved in that, and because we lived so far away from the rest of the Chinese community.

Kaylin therefore asserted that her unmarried status was a result of her 'serious' life and physical isolation from the rest of the Chinese community in Sydney (a point I discuss in more detail in Chapter 5). The combination of these two factors, according to Kaylin, meant that the opportunity to meet a potential husband was limited. Implicit in Kaylin's reasoning is the understanding that marriage was only possible with a Chinese Australian man. When I asked Kaylin to confirm that point, she responded in the affirmative: 'Yeah. I would say so because we never mixed. We just stayed at home and did our studies and our chores around the house'. Kaylin's experiences therefore reiterate the emphasis on endogamy within Chinese Australian families.

Given that scholarship has almost always constructed Chinese women as (absent) wives who were dependant on their husbands in terms of financial security and/or migration, it is important that the presence of 'single' Chinese Australian women is acknowledged. Their presence indicates an alternative subject position that was not based on a role within the institution of marriage, and in Mei-Lin and Margaret's instances, indicates a type of agency that has not been examined elsewhere. Their position as unmarried women with various living arrangements provides further evidence for Bagnall's (2006) assertion that family life existed in Chinese Australian communities in ways that did not conform to western conceptions of the nuclear family unit. Furthermore, given Choi's (1975) argument that Chinese men were 'forced' to find Chinese wives overseas due to the lack of unmarried Chinese women in Australia during this period, it is unclear why the census records indicate such a high proportion of unmarried Chinese women in Australia and why their presence has been so neglected in the literature.

Presence of Australian-born female children

Emphasis on the 'bachelor' society of Chinese Australian men and their (absent) wives has also failed to recognise the cohort of young daughters—whether of 'full' or 'mixed' Chinese descent—who were present in Chinese Australian families throughout the White Australia Policy era (with the exception of Loh and Ramsey 1986 and Tan 2003). An examination of census data provides visibility to this cohort of young daughters. Between 1911 and 1954, the clear majority of Chinese females were in the 0–14 years of age bracket, representing between 31.5 per cent (in 1933) and 47.3 per cent (in 1911) of the total Chinese female population (see Table 3.5). Of the 19 interview participants, 17 spent their childhood and/or teenage years in Australia. This group includes all 13 of the Australia-born participants who spent their formative years living in Australia, as well as Patricia who migrated at six years of age, and Helen, Sandy and Mary who came to Australia from Hong Kong as teenagers to attend high school. Childhood photographs of participants help reveal this presence of young daughters in the White Australia context. For example, a photograph of Mei-Lin and her cousin in Texas, QLD in the 1950s (Figure 3.1), and Mabel and her sisters in Merrylands, Sydney c.1945 (Figure 3.2) provide some indication of the young female cohort that were present in Australia throughout the period. Experiences of growing up in White Australia, particularly of the fourteen participants' who spent their childhood years in Australia, will be explored in the following chapters.

Table 3.5 Age Groups of Chinese Australian Females, 1911–1966

Age group	1911 (%)	1921 (%)	1933 (%)	1947 (%)	1954 (%)	1961 (%)
0–14	1,135	1,139	987	1,089	1,534	2,162
	(47.3)	(39.1)	(31.5)	(27.9)	(30.7)	(28.1)
15–24	518	636	713	682	1,019	2,269
	(21.6)	(21.8)	(22.7)	(17.5)	(20.4)	(29.5)
25–34	367	510	534	648	697	1,005
	(15.3)	(17.5)	(17.0)	(16.6)	(13.9)	(13.1)
35–44	188	377	374	644	711	772
	(7.8)	(12.9)	(11.9)	(16.5)	(14.2)	(10.0)
45–54	75	164	295	361	512	718
	(3.1)	(5.6)	(9.4)	(9.3)	(10.2)	(9.3)
55+	13	83	222	450	531	757
	(0.5)	(2.8)	(7.1)	(11.5)	(10.6)	(9.9)
Unspecified Adults	39 (1.6)	–	–	–	–	–
Unspecified Children	61 (2.5)	–	–	–	–	–
Not Stated	2	7	12	27	–	–
	(0.1)	(0.2)	(0.4)	(0.7)	–	–
TOTAL	2,398	2,916	3,137	3,901	5,004	7,683
	(100)	(100)	(100)	(100)	(100)	(100)

Source: CBCS (1917, 1924, 1937, 1951, 1958, 1964).

Figure 3.1 Mei-Lin Yum and cousin Coral in Texas, Queensland, c.1950. This photograph of Mei-Lin (right) and her cousin Coral (left) was taken on Christmas Day. Mei-Lin was born in 1942 in Texas, QLD and was one of two children (she and a younger brother) born to Australian-born parents.

Source: Mei-Lin Yum (private collection).

Figure 3.2 Mabel Lee and her sisters in Merrylands, Sydney c.1945. Australian-born Mabel (right) poses with her sisters Hazel (left) and Judy (centre) in Merrylands where the family ran a profitable fruit shop. Mabel was born in 1939 in Warialda, a country town in Northern NSW. She also has four brothers.

Source: Mabel Lee (private collection).

Uncovering an enduring presence

The presence of Chinese Australian females in the White Australia period was not momentary or sporadic. As data on population numbers indicate, Chinese Australian females were present in the nation from 1901 and increasingly so throughout the period. Chinese Australian females were also living in the nation in the nineteenth century, although in much smaller numbers—two women were recorded in the colony of New South Wales in 1861 and an additional eight in Victoria. By 1881, 259 Chinese Australian women were recorded in Australia (Bagnall 2011). Census data on the length of residence of foreign-born Chinese Australian females provide a picture of their long and enduring presence in the nation. For example, in 1911, 76 Chinese migrant females had been resident in the country for over 20 years, meaning they had migrated to Australia prior to the 1890s (see Table 3.6). The proportion of long-term migrant residents (those that were resident for more than 20 years) increased significantly prior to the onset of World War II (reaching 52 per cent of the migrant female Chinese population in 1933). However, in the post-war period, the proportion of long-term residents declined rapidly as new immigrant arrivals increased. This is reflected in the large increase in migrant females who had been resident in Australia for less than 20 years between 1933 and 1947, and the increasing number of women falling into that category up until 1961.

Table 3.6 Length of Female Chinese Migrant Residence in Australia, 1911–1961

Year	Under 20 years[1]			20+ years[2]			Unspecified		
	Full Chinese	*Mixed Chinese*	*Total*	*Full Chinese*	*Mixed Chinese*	*Total*	*Full Chinese*	*Mixed Chinese*	*Total*
1911	141	4	145	74	2	76	45	2	47
1921	–	–	–	–	–	–	–	–	–
1933	92	4	96	111	9	120	16	1	17
1947	601	40	641	102	13	115	43	2	45
1954	1,190	53	1243	278	24	302	38	0	38
1961	3,150	132	3282	278	13	291	117	3	120

1 1954: Under 15 years; 1961: Under 21 years.
2 1954: 15 years and over; 1961: 21 years and over.

Source: CBCS (1917, 1924, 1937, 1951, 1958, 1964).

The longstanding presence of Chinese Australian women in the period is also reflected in the multiple generations of females who were resident in the nation. The family histories of Australian-born interview participants provide important evidence of this generational imprint as several were descendants of immigrant mothers, while others had mothers, and even grandmothers, who were also Australian-born. For example, Mei-Lin's maternal great-grandmother was a migrant, arriving in Australia in the nineteenth century. Her maternal grandmother (Agnes Mary; see Figure 3.3) and mother were subsequently Australian-born. She explained:

> She [Agnes Mary] was born here in Australia with an Australian father [...]. Her husband was Chinese. His name was Tart Lumbiew. My mother's maiden name was Lumbiew and my mother was one of four children. She was born, I think, in Surry Hills and then they moved to Kensington and they were educated in Sydney.
>
> (Mei-Lin)

Figure 3.3 provides a further (visual) illustration of the first two Australian-born generations of women on Mei-Lin's maternal line.

Figure 3.3 Two generations of the Lumbiew family, Sydney, c.1900. Australian-born Agnes Mary and Tart Lumbiew are pictured with their first three (of four) children, one of which is Mei-Lin's mother (far right). Agnes Mary and Tart were married in Sydney at approximately 20 years of age and 40 years of age, respectively.

Source: Mei-Lin Yum (private collection).

Edith similarly comes from a long line of Chinese Australian women. Edith is third generation[4] Australian as both her parents were born in Australia to Chinese migrants who arrived in the colonies in the nineteenth century. Edith recalled:

> My mother's family were on the South Coast [of New South Wales]. I know my eldest aunty was born in Narrandera then my mother was born in Nowra. Her younger sister was born in Milton and the youngest son was born in Port Kembla so he had market gardens in those areas.

Marina is also third generation Australian. Her paternal grandparents migrated to Australia from Zhongshan, China in the late nineteenth century (c.1892) while her maternal grandmother was an English woman who had three children with a Chinese Australian man. Marina's mother, Ruby Wong Chee, was therefore of Chinese-Anglo descent and was born in Sydney in 1897. The family portrait of the Fay family (Figure 3.4), which includes Marina's paternal grandmother, parents, Marina and her siblings, and one nephew, provides a pertinent visual illustration of the long and enduring presence (and family legacy) of Chinese Australian females. The extended Fay family portrait (Figure 3.5) emphasises this female presence given the striking number of female family members across four generations—15 females and 14 males.

Figure 3.4 Four generations of the Fay family, 1941. As the handwritten caption underneath this photograph indicates, this image of the Fay family was taken at the 25th wedding anniversary of Harry and Ruby Fay née Wong Chee (Marina's parents). Marina's paternal grandmother is seated in the centre with her son Harry and daughter-in-law, Ruby, seated on either side. Harry and Ruby's children surround them (seven-year-old Marina standing between her mother and grandmother in the front row), and Harry and Ruby's grandson 'Baby Brian' is held in the arms of Charles Cum (centre back). Marina's paternal grandmother migrated to Australia with her husband in the late 1800s.

Source: Marina Mar (private collection).

Figure 3.5 The extended Fay family, c. 1940. Much like the family portrait of Marina's immediate family (Figure 3.4), this image of the extended Fay family depicts at least four generations. The number of female members in the extended family is striking in this image and provides further visual evidence of the long and enduring presence in Australia of women in the Fay family.

Source: Marina Mar (private collection).

Kaylin's ancestry similarly highlights the extensive presence of Chinese Australian females. While Kaylin's father was a Chinese migrant—arriving in Sydney in 1900—her mother was Australian-born, being one of six children of migrant parents. Kaylin's maternal grandfather arrived in Australia in 1868 to establish a new life on the Victorian goldfields, while her grandmother arrived later as his wife. Kaylin explained:

> He [grandfather] did return to China to marry and he came back with his new bride who had been brought up in a German mission school in Hong Kong. And they worked among the Chinese community here in Sydney for a number of years, with the Presbyterian Church. […] Grandfather and Grandmother had six children—three sons and three daughters.

Like Kaylin, Nancy is also second generation Australian. Her father was Australian-born and her mother was a Chinese migrant. Her paternal grandparents, Yit Quay Loong (grandmother) and Yuen Mu Loong (grandfather), were both migrants who arrived in Australia around the turn of the twentieth century. They were married in Melbourne at the age of 19 and 36, respectively. Nancy's father, Henry Loong, was their eldest son and was born in 1908/1909. The Loong family portrait (Figure 3.6) highlights the female presence in Nancy's paternal

Figure 3.6 The Loong family, Melbourne, c. 1917. This photograph of Nancy's paternal
grandparents, Yit Quay Loong (seated left) and Yuen Mu Loong (seated right),
with their first five children, was taken before the family returned to China.
Nancy's father, Henry (standing back centre), was their eldest son. The other
children (from left to right) are Marjorie, Florence, George and Grace.

Source: Nancy Buggy (private collection).

line through the inclusion of Nancy's grandmother, Yit Quay, and her daughters—Nancy's paternal aunts.

The presence of Australian-born women like Mei-Lin, Edith, Marina, Kaylin,
Nancy and their Australian-born or migrant mothers, grandmothers and even
great-grandmothers indicate the longstanding presence of females in Chinese
Australian communities as far back as the 1870s. Their presence also indicates
the crucial role females played in the longevity and continuation of Chinese
Australian communities in the nineteenth and twentieth centuries.

International migration and mobility

In Chapter 1, I briefly detailed the common conceptualisation of males as primary
actors in the migration process that has been propagated by research that focuses on
political and economic reasons for male migration as uncovered by the 'push–pull'
approach, and partly as a result of the official patriarchy manifest in government
records of migration flows. I explained that such gendered assumptions of migration, in records and research, have assumed the dependence of women on the male
migrant—either staying behind or passively accompanying their independently

mobile husband. Consequently, women have been rendered invisible in international migration stories (as argued by Ryan 2003 among others). I also highlighted how such 'gender blindness' in migration paradigms, in addition to generalised assumptions of women's positions in the 'traditional' Confucian family system, have particularly influenced understandings of Australia's Chinese past and subsequently rendered Chinese Australian females invisible. This invisibility has not been confined to the Australian context. 'Modern' Chinese migration across the globe (between 1842 and 1949 as defined by McKeown 1999) has more broadly been defined as male-centric and it is within this broader patriarchal migration paradigm that the gender-blindness in Australia's Chinese history must be viewed.

The birthplace statistics I presented earlier in this chapter are significant as they challenge these dominant understandings of modern Chinese migrations, particularly to Australia, and female immobility. For example, the large numbers of 'foreign-born' Chinese Australian females indicate that Chinese women, like their male counterparts, were mobile and crossing international borders. This was so much the case in Australia that female mobility and immigration accounted for the majority of female Chinese Australian population growth in the post-war era (as was similarly evident in New Zealand, see Ip 1995). It is difficult, in this instance, to use specific migration data to understand the population growth of Chinese Australian females as it is not clear which arrivals and departures were made by Australian-born or foreign-born individuals and permanent/temporary losses or gains in any given year are not indicated (as discussed by Choi 1975: 44). The post-war migration data are also compromised as arrivals/departures according to 'race' are not available—rather 'nationality' figures were used[5]. It is for this reason that I have utilised birthplace census data as a more reliable source for understanding increases in the female Chinese population.

Migration data do, however, provide other important information that supports my arguments of female mobility. Migration figures from 1914 to 1965 as collated by Choi (1975) indicate constant flows of Chinese in and out of the country between those years. In fact, in the post-war years, both male and female Chinese migrant arrivals far exceeded departures so that the male population had a net gain of 4,536 individuals and the female population 2,252 individuals (Choi 1975: 62; Table 3.7b). It is also important to note that in the earlier period between 1914 and 1947, female Chinese arrivals outnumbered departures by 321 compared to a loss of 8,060 Chinese men (Table 3.7a). Choi (1975) stated that this increase in females between 1914 and 1947 was not 'appreciable' (p: 45). In contrast, I argue, that these data as well as the post-war gains indicate mobility and presence of Chinese females in the nation which should be acknowledged especially given decreases in the male population. Such movements challenge patriarchal assumptions of female immobility as well as dominant understandings of female absence in this historical and geographical context.

Table 3.7a Chinese Arrivals and Departures to/from Australia, 1914–1947[1]

Year	Male			Female		
	Arrivals	Departures	Gain (+) Loss (−)	Arrivals	Departures	Gain (+) Loss (−)
1914	3,465	3,869	−404	107	81	+26
1915	3,362	4,233	−871	87	118	−31
1916	3,361	3,288	+73	80	113	−33
1917	2,836	2,602	+234	86	110	−24
1918	2,367	2,265	+102	93	93	0
1919	2,415	2,638	−223	103	108	−5
1920	3,323	4,045	−722	191	170	+21
1921	3,426	4,634	−1 208	204	220	−16
1922	3,417	3,522	−105	196	203	−7
1923	3,699	3,509	+190	165	153	+12
1924	3,237	3,291	−54	173	136	+37
1925	2,471	2,768	−297	145	132	+13
1926	3,162	3,521	−359	151	189	−38
1927	3,259	3,886	−627	199	221	−22
1928	3,065	3,257	−192	242	205	+37
1929	2,693	2,990	−297	220	230	−10
1930	2,461	2,977	−516	197	250	−53
1931	2,062	2,540	−478	246	269	−23
1932	1,569	1,928	−332	204	201	+3
1933	1,321	1,730	−409	162	181	−19
1934	1,399	1,431	−32	181	182	−1
1935	1,351	1,435	−84	187	152	+35
1936	1,458	1,544	−86	207	208	−1
1937	1,377	1,409	−32	239	159	+80
1938	1,462	1,271	+191	316	204	+112
1939	1,115	1,161	−46	524	411	+113
1940	778	1,003	−225	326	293	+63
1941–3	Not available	Not available	Not available	Not available	Not available	Not available
1944	25	414	−389	15	17	−2
1945	113	38	+75	36	12	+24
1946	326	680	−354	126	88	+38
1947	772	1,311	−589	191	199	−8
Total	67,124	75,190	−8,066	5,599	5,278	+321

1 The reason for presenting data for 1914–1947 separately from 1949–1965 is not stated in Choi (1975). However, Choi (1975: 41) does explain that 1947 was the year in which conditions of each exemption category under the *Immigration Restriction Act* were clarified. In addition, with heavy restrictions on Chinese migration (via exemption categories) up until 1947, the census of that year counted the lowest number of Chinese Australians in the entire White Australia Policy era (see Table 3.1).

Source: Adapted from Choi (1975).

Table 3.7b Chinese Arrivals and Departures to/from Australia, 1949–1965[1]

Year	Male			Female		
	Arrivals	Departures	Gain (+) Loss (−)	Arrivals	Departures	Gain (+) Loss (−)
1949–51	959	612	+347	338	255	+83
1950–2	855	413	+442	358	233	+125
1951–3	724	429	+295	382	267	+115
1952–4	667	471	+196	368	274	+94
1953–5	608	526	+82	351	284	+67
1954–6	722	585	+137	387	298	+89
1955–7	848	702	+146	454	346	+108
1956–8	980	751	+229	488	351	+137
1957–9	1,051	729	+322	518	347	+171
1958–60	1,136	657	+479	535	306	+229
1959–61	1,319	697	+622	587	303	+284
1960–2	1,397	776	+621	590	299	+291
1961–3	1,373	959	+414	584	355	+229
1962–4	1,296	1,074	+222	538	401	+137
1963–5	1,155	1,173	−18	513	420	+93
Total	15,090	10,554	+4,536	6,991	4,739	+2,252

1 There is no explanation in Choi (1975) regarding the lack of data for 1948.

Source: Adapted from Choi (1975).

Given the presence of foreign-born females in White Australia and the flows of Chinese females in and out of the country between 1914 and 1965, interesting and important questions emerge regarding the nature of mobility and migrations of Chinese women to Australia. For example, did entries into the nation reflect temporary settlements? Did the experiences of Chinese women who migrated to Australia follow dominant understandings of passive wives following their husbands abroad? Or did their movements challenge patriarchal power structures within Chinese family systems? Census and migration data are limited in their ability to give insight into these questions as motivations and reasons for migration were not documented. As I will detail below via the discussion of migrant wives and international mobility of Australian-born participants, first-hand accounts shine some light on the nature of Chinese mobility in the White Australia period and the complexity of the female migrant experience[6].

Migrant wives

As touched on above, it was common for mothers or grandmothers of Australian-born Chinese to have arrived in Australia as migrant wives—being married in China and then migrating to Australia with their husbands, or following them later. Stella's family history, which opened the introduction to this book, provides a pertinent illustration of this type of movement, as does Nancy's:

> [Dad] never actually lived in China. He commuted. He would come to Sydney to work and then go back to visit and that's how he met Mum. And after they were married she went to live with his family [...] in one of his trips back to her, she pleaded with him to take her to Australia with him. [...] So I think it was about 1932.
>
> (Nancy)

Unlike Nancy and Stella, Patricia was not Australian-born, but a migrant herself. She arrived in Surry Hills, Sydney, in 1947 (after World War II and just prior to the establishment of the People's Republic of China) at the age of six with her family. Patricia's father had arrived years earlier, being sponsored by his brother-in-law to help in their restaurant Nan King Café in Campbell Street. This type of sponsored migration of Chinese men by Chinese Australian entrepreneurs (whom they were often connected to by kinship, village or clan ties) was common in the White Australia Policy period, particularly after legislation reforms in 1934 (Choi 1975: 41). According to Patricia, the earlier migration of her father was also influenced by the patriarchal Confucian family system. She explained:

> In those days, being the eldest son, if you had a job, you had to help the family. So of course his father had other sons and daughters and needed help [as] the business back home wasn't doing well.

It seems that Patricia's father's migration was first and foremost rooted in his sense of Confucian 'family duty' and the need to financially support his family in China—a common practice among Chinese men in the nineteenth- and early twentieth centuries. Patricia went on to explain that once in Australia, 'He missed the family, so we came out. That's why I came'. After a long and strenuous boat trip from Dongguan in Guangdong province to Australia, Patricia's family settled in a rented house in inner-city Sydney—a 'big house' that Patricia described as 'wonderful compared to what I lived in back in China'. Thus, while Patricia's migration story reveals the entry of female Chinese into White Australia (herself and her mother), Patricia and her mother's movement was dependent on Patricia's father—Patricia was positioned as a dependent child while her mother, having stayed in China to care for the family, later followed her husband. In this way, Patricia's migration story closely follows gendered understandings of migration and parallels the migration experiences of many overseas Chinese women in other 'Gold Mountain' countries (see, for example, Ip 1990 and Ling 2000).

Often precariously positioned as temporary entrants, relaxations to naturalisation and immigration restrictions after 1956 allowed many of these migrant wives to settle permanently in Australia. For example, Eileen's mother, Fong See Lee Yan née Wai Hing, migrated from Canton to Australia in the early decades of the twentieth century as the wife of a Chinese herbalist. After her first husband's death, she married Eileen's Australian-born father, Frederick Lee Yan, in 1945. Despite this marriage, Eileen's mother was under constant threat of being deported until her naturalisation in the early 1960s. Eileen recalled:

> Mum was naturalised [on the] 1st of March 1963. Then that gave her the right to stay. That's what I couldn't understand, because Dad being born here and he's got papers to say he's an Aussie. [...] I thought well how come Dad being an Aussie didn't give Mum the right to stay? They said no. So back in those days they were still going to send my mother back to China, even though she married an Australian-born Chinese.

Australian-born Doreen similarly recalled the circumstances that allowed her Chinese mother to stay in Australia despite strict immigration restrictions. Remaining in China as the second wife of a sojourning market gardener, Doreen's mother managed to obtain a one-year temporary entry visa for herself and her first born child (Doreen's elder brother) to reunite with her husband in Australia in the late 1930s:

> So after the child was born, my mother decided that she would come and visit my father because that was all the White Australia Policy allowed them. Again, she had bought somebody's papers because those immigration documents put her about ten years older than what she actually is. So then, she comes out here, becomes pregnant with me, World War II Pacific action flares up and she can't get back to China.
>
> (Doreen)

Given the unstable political situation and the outbreak of civil war in China after World War II, Doreen's father, who had always had the belief that he and his family would return to China, decided that 'There was a better life for his children here' (Doreen) and consequently gave up his plight to return 'home'. In Doreen's memory, however, this time was always marked by uncertainty as her parents were unable to be naturalised:

> Well, I mean it was very difficult because I can remember as I said, my mother came out here on just obviously a temporary entry and she had to stay because she couldn't get back to China, and having to go into immigration every year to get the extension, and even my father had to do that but my mother more so because she was supposed to be on a more temporary status than what my father was. So there was a period where they kept being frightened they would be forced to return to China.

Despite the uncertain situation, Doreen's family found themselves in as a consequence of immigration restrictions, their experiences highlight an important form of agency. Firstly, Doreen's mother 'decided' that she would visit her husband in Australia. This active decision challenges dominant assumptions of women as passive migrants. Secondly, by purchasing documents and claiming another identity, Doreen's mother 'worked the system' and found a weakness in the restrictive immigration legislation (an initiative that was not uncommon). It was through this scheme that she was able to enter Australia and, with changes to naturalisation legislation in the 1950s, eventually become an Australian citizen along with her husband and son (who was born in China) in 1958–59. In this way, Doreen's mother's experience indicates that some Chinese women occupied more empowered positions in the migration process—being able to negotiate and work around immigration restrictions rather than being passive victims of discriminatory policy.

Two of the interview participants—Lily and Daphne—were themselves 'migrant wives'; however, their experiences did not follow the traditional or expected route of passive or dependant housewives following their migrant husbands. For example, Lily arrived in Australia from Hong Kong in 1971 (at 32 years of age) with her husband and children. Both Lily and her husband wished to leave Hong Kong due to the impacts of the Cultural Revolution in China and the possibility of a better life and education for their children. She explained:

> I was in Hong Kong working at the Hong Kong University Library then. Even Hong Kong was affected [by the Cultural Revolution in China] and so we were thinking of leaving Hong Kong, not just because of the political situation, but also my husband and I felt that Hong Kong was not an ideal place to raise children because, you know, of the educational system and the crowded sort of living quarters.
>
> (Lily)

On her and her family's arrival in Sydney, Lily quickly obtained professional employment with the help of her master's degree and specialised training:

> I actually got two offers when I started to look for jobs: one is just librarian and the other one was the Oriental Librarian at the University of Sydney. They had a special collection of Chinese and Japanese [texts] and they need[ed] someone with library qualifications as well as, you know, able to read Chinese, so I fitted in.

Lily's recollections indicated that she was not positioned as a passive wife following her husband abroad, but was actively involved in the migration process—from the initial decision to migrate through to the maintenance of the economic stability of her family in their new home. This position challenged Confucian expectations of the submissive wife and mother. Lily's experiences of mobility and contribution to Australia's cultural diversity also deviate from male-oriented understandings of migration as well as androcentric (and ethnocentric) assumptions of Australian national development.

Indicating further diversity of female migrant experience in the post-war period, New Zealand-born Daphne migrated to Australia from New Zealand in 1964 at the age of 25. Her move across the Tasman Sea was a result of her marriage to a Chinese Australian resident in Sydney: 'I got married and then came here because my first husband was from here' (Daphne). Unlike Lily who seemed to have quite a problem-free entry into Australia, Daphne had firsthand experience of the discriminatory practices associated with Australian immigration policy at the time. She recalled the measures she had to go to in order to be permitted into the country in 1964 despite her New Zealand citizenship:

> Would you believe, the day I was to fly out of Wellington to come to Australia I got a call from the airline to say, 'Oh, look we've just realised you need a visa'. And I said, 'Oh, well…I checked. I didn't need one; I'm on a New Zealand passport'. […] This is White Australia Policy, right? […] I'd travelled…been back to China and everything, done all these things and never ever had it hit me, but it really hit me when I came.[…] Anyhow, I needed a visa and I only found out the reason I needed the visa was because I was Chinese.

Daphne's experiences of the Australian immigration system reflects the contested nature of her right to 'belong' in New Zealand as well as Australia on account of her 'Chineseness'. That is, her 'Chineseness' restricted her citizenship rights to move freely between New Zealand and Australia. Daphne was not alone in experiencing this type of discrimination. Gittins (1987) has also described the way in which the 'fifty per cent of Asian blood' in her veins meant that her 'irregular entry' into Australia was reported to the Immigration Department and she was required to report to them at regular intervals. Such an encounter with an immigration official on her landing in Australia from Hong Kong made her 'feel like a common criminal on parole' despite the fact that she held a British passport all her life (p: 113–14).

International mobility of Australian-born females

Interviews with Chinese Australian women also indicated that females born in Australia were internationally mobile. The Australian-born mothers of Marina and Stella travelled to and from China or Hong Kong on multiple occasions. Marina's mother, Ruby Wong Chee, was born in Sydney to an English mother and Chinese father and, after being adopted into a Chinese family and spending some time in Glen Innes, New South Wales, was sent to China where she later married a Chinese Australian, Harry Fay. Marina explained:

> […] he married my mother in 1916. He had to go to China and I think they married in Zhongshan somewhere, in the Canton area. Then they came back to Australia, and they came on the boat three months—because she'd had a bit of smallpox before that, she'd arrived—so when they got here they had to go to the quarantine station for so many months, I suppose. Then they went straight to Inverell, and they lived on top of the shop.

Therefore, as explained by Wilton (2004), although Ruby Wong Chee and Harry Fay travelled to China to marry, their future was in Australia. As presented in the opening vignette to this book, Stella's Australian-born mother was also sent back 'home' to be married:

> [My mother] didn't get married until she was about twenty-one and they decided that she was getting really old, so they arranged for her to go to Hong Kong and marry a friend's son, which she did.
>
> (Stella)

Unlike Ruby, however, Stella's mother returned to Australia alone: 'she left him to come back to Thursday Island to have me after her first baby died'. Stella's mother also made an additional two trips back to Hong Kong, the first time to search for her estranged husband: 'when she went back to find him, my father, he was gone and she never found out where he went to', and the second time to obtain medical treatment for her son: 'when she met another man and had my first half-brother, she went back to Hong Kong with him because he was sick and he wanted some traditional Chinese medicine'.

Ip (2002) has argued that the repatriation of young overseas Chinese females to China was quite a common experience in the early decades of the twentieth century. While boys were sent back to China to receive a 'proper' Chinese education, girls were sent 'home' to find 'proper marriage partners' (Ip 2002: 155), as was the case for Marina's and Stella's mothers. Many of the interview participants explained that marriage to a non-Chinese was totally discouraged by parents. As Ina recalled: 'I remember most of our parents would say, "Marry a Chinese. Don't marry an Australian"'. As discussed in the previous chapter, marriages were often arranged between locally born individuals, and the repatriation of Chinese Australian men for the purpose of marrying back in their home village was common. Sending daughters back to China was perhaps another way of ensuring the continuation of endogamy in the Chinese Australian context. It is interesting to note that Marina's mother and father were both Australian-born yet their marriage took place in China. In the New Zealand context, these types of marriages between two local-born Chinese back in China were common (Ip 2002). More research is needed to further examine the extent of these unions among early twentieth-century Chinese Australians.

Two of the Australian-born interview participants were themselves internationally mobile in the White Australia period—leaving Australia to live abroad for some time. Ina left Australia in 1968 with her husband and did not return to Australia until 1989, well after the official abandonment of the White Australia Policy. Her life abroad was anything but static, comprised of several international relocations over the 21-year period. She explained her mobility in the following way:

> Now the thing in those days was to go overseas for a couple of years. [...] I think we had three years in Amsterdam and then stayed in London for six years and the children were born there. And after six years, London was in the pits of depression and a friend of ours [...] said, 'Oh, you know you can go and be an architect in Hong Kong. They are looking for expatriate architects'. And so we applied for that and it took a year to come through. And we got through to Hong Kong and we lived there for twelve years.

(Ina)

While Ina and her husband's eventual settlement in Hong Kong was driven by her husband's profession as an architect, Ina's entire migration experience was not shaped by her position as a 'dependant wife'. Their time spent in Amsterdam was, in fact, largely shaped by Ina's employment:

> [My husband's] Sydney degree wasn't recognised in Holland, and he worked as a draftsman [...] so [in the summer holidays] he would leave his job, and I would keep my job at the International School and come back in September when the term opens.

While Ina moved abroad with her husband, Nancy on the other hand set sail to Canada independently after completing her nursing qualifications in 1963. She explained:

> I did my four years nursing at Royal Prince Alfred. Did a year over in Queen Elizabeth in Adelaide—my midwifery—and then about six weeks later I was on the ship going to Canada.

It was in Canada that she met her Australian husband and lived for two years before settling in London for an additional year and returning to Australia in 1966. Once again, further investigation is needed to determine whether or not Nancy's initial independent travel was unique among her cohort. What can be said, however, is that her mobility illustrates deviance from the female 'norm' of dependence or immobility perpetuated by male-centred migration paradigms.

Geographical distribution and intra-national mobility

Census data indicate that the long and enduring settlement of Chinese Australian females in the White Australia Policy era was not confined to a select few towns or major cities. Rather, the population of Chinese Australian females was geographically dispersed, with females recorded in all states and territories (with the exception of the Australian Capital Territory), in all census years between 1911 and 1966. Therefore, despite assumptions of their absence, Chinese Australian females contributed to populations across the continent throughout the White Australia period. However, at the level of states and territories, the geographical distribution of Chinese Australian females across the nation was not even. Female Chinese Australians were predominantly located in New South Wales (NSW), Victoria (VIC) and Queensland (QLD) (see Table 3.8); with NSW almost always having the national majority (the exception was in 1921 when the VIC population outnumbered the NSW population by 452). For example, in 1911, 35.7 per cent of the total Chinese female population were counted in NSW. At the census of 1966, this had increased to 51.5 per cent. Not surprisingly, the predominance of females in the three eastern states mirrored distribution patterns of the broader Chinese Australian community and the total Australian population. NSW, QLD and VIC consistently accounted for more than 80 per cent of the total Australian population as well as the total Chinese Australian population (Choi 1975: 51; see also Jones 2005). The concentration of Chinese Australians in the eastern states reflected the longer histories of Chinese settlement (all three attracting large numbers of Chinese immigrants during the gold-rush period), and the position of the three eastern states as the economic, political and cultural centres of nineteenth- and twentieth-century Australia.

Table 3.8 State Distributions of Chinese Australian Females, 1911–1966

State	Population at Census (%)						
	1911	*1921*	*1933*	*1947*	*1954*	*1961*	*1966*
NSW	855	1,058	1,218	1,886	–	3,811	4,935
	(35.7)	(36.3)	(38.9)	(49.1)		(49.7)	(51.5)
VIC	645	1,510	760	750	–	1,735	1,428
	(26.9)	(51.8)	(24.3)	(19.5)		(22.6)	(14.9)
QLD	576	782	794	904	–	1,348	1,281
	(24.0)	(26.8)	(25.4)	(23.5)		(17.6)	(13.4)
SA	68	79	50	39	–	123	230
	(2.8)	(2.7)	(1.6)	(1.0)		(1.6)	(2.4)
WA	64	95	118	141	–	274	389
	(2.7)	(3.3)	(3.8)	(3.7)		(3.6)	(4.1)
TAS	79	38	37	30	–	84	120
	(3.3)	(1.3)	(1.1)	(0.8)		(1.1)	(1.3)
NT	111	116	154	126	–	277	320
	(4.6)	(4.0)	(4.9)	(3.3)		(3.6)	(3.3)
ACT	0	0	0	1	–	20	78
				(0.03)		(0.3)	(0.8)
Australia	2,398	2,914	3,131	3,898	5,002	7,672	9,581
	(100)	(100)	(100)	(100)	(100)	(100)	(100)

Source: CBCS (1917, 1924, 1937, 1951, 1958, 1964, 1971).

In addition to the settlement of Chinese Australian females in the eastern states, the census data indicate a relatively high concentration of Chinese Australian females in the Northern Territory (NT). In regards to the total Australian population, the population of the NT remained quite low throughout the White Australia period. In 1911, the population of the NT (3,310 persons) made up only 0.07 per cent of the total Australian population (4,455,005 persons). By 1966, the population of the NT had increased to 56,504 but continued to comprise a mere 0.5 per cent of the total Australian population (11,599,498; CBCS 1971; Table 3.8). As has been noted elsewhere (see, for example, Jones 2005), within this relatively small population, Chinese Australians made up a proportionally significant number. Prior to Federation in 1901, the NT was home to Chinese 'coolies' who were contracted to work the goldfields at Pine Creek, and later, to provide cheap labour for the construction of the railway line between Pine Creek and Palmerston. In 1878, Chinese were the largest non-Aboriginal group in the Territory and their numerical dominance continued into the twentieth century. As Jones (2005: 16) has argued, for the first decade following 1901, the Chinese Australian population in the NT outnumbered Europeans. Census data indicate that in 1911, Chinese Australians comprised 40.5 per cent (1,339 of 3,310) of the total NT population. By 1966, this had fallen to 1.2 per cent (672 of 56,504). Thus, while the total population of the NT increased by over 53,000 people between 1911 and 1966, the Chinese Australian population of the territory almost halved.

Despite this large statistical decrease, the population of Chinese Australians in the NT throughout the White Australia period can be seen to be numerically significant when we compare it to the total Chinese Australian population across the country. For example, in 1911, 5.2 per cent of Australia's Chinese population were resident in the NT, dropping to 2.5 per cent by 1966 (CBCS 1971). More particularly, in regards to the female Chinese Australian population, as Table 3.6 indicates, the NT female Chinese Australian population comprised 4.6 per cent of the total female Chinese Australian population in 1911, and 3.3 per cent by 1966. Perhaps more importantly, in 1911, the Chinese Australian female population comprised 3.4 per cent of the total NT population (111 of 3,310), and 19.3 per cent of the total female population of the NT (111 of 576). Despite falling to 0.9 per cent of the NT population (320 of 37,433) and 2 per cent of the total female population of the NT (320 of 15,925) in 1966, a strong presence of Chinese Australian females in the NT remained clear. The reason for this remains unclear.

Within the states and territories, data from 1911 to 1966 indicate a presence of Chinese Australian females in both urban and rural areas, however, settlement predominated in urban locales[7]. For example, at the national level, in 1921, 71.7 per cent of Chinese females in Australia lived in urban centres compared to the 28.3 per cent who lived in rural areas. This uneven distribution continued throughout the period. In fact, there was an overall increase in the urban female Chinese Australian population between census years, with rapid increases occurring between 1947 and 1954 (see Table 3.9). Rural locations, on the other hand, experienced steady losses. Indeed during this period, the Australian population as a whole was experiencing massive urbanisation. However, census data indicate that Chinese Australian females were more urbanised than the national population. For example, while 71.7 per cent of Chinese Australian females lived in urban centres in 1921, 62.1 per cent of the total Australian population resided in urban locales. In 1954, the gap widened with 93.3 per cent of Chinese Australian women living in urban areas compared to 78.7 per cent of the total Australian population. By 1966, 96.1 per cent of Chinese Australian females were resident in urban locations compared to 82.9 per cent of the total Australian population.

Table 3.9 Rural/Urban Distributions of Female Chinese Australians, 1911–1966

State		Census Year						
		1911[1]	*1921*	*1933*	*1947*	*1954*	*1961*	*1966*
NSW	% Urban	44.8	84.4	82.7	90.1	–	96.7	97.5
	% Rural	55.2	15.6	17.3	9.9	–	3.3	2.5
VIC	% Urban	38.8	38.2	78.7	88.3	–	99.1	96.2
	% Rural	31.2	61.8	21.3	11.7	–	0.9	3.8
QLD	% Urban	16.1	51.0	45.3	56.3	–	85.9	89.
	% Rural	83.9	49.0	54.7	43.7	–	14.1	10.9
SA	% Urban	66.2	69.6	82.0	82.1	–	97.1	95.7
	% Rural	33.8	30.4	18.0	17.9	–	2.9	4.3
WA	% Urban	56.3	68.4	61.9	68.1	–	83.9	93.8
	% Rural	43.9	31.6	38.1	31.9		16.1	6.2
TAS	% Urban	11.4	42.1	56.8	73.3	–	96.4	94.2
	% Rural	88.6	57.9	43.2	26.7	–	3.6	5.8
NT	% Urban	0.0	72.4	80.5	53.2	–	92.4	93.4
	% Rural	100.0	27.6	19.5	46.8	–	7.6	6.6
ACT	% Urban	0.0	0.0	0.0	100.0	–	100.0	98.7
	% Rural	0.0	0.0	0.0	0.0	–	0.0	1.3
Australia	% Urban	34.0	71.7	71.0	79.8	93.3	94.8	96.1
	% Rural	66.0	28.3	29.0	20.2	6.7	5.2	3.9

1 In 1911, 'Urban' is classified as residing in the capital city or its suburbs only.

Source: CBCS (1917, 1924, 1937, 1951, 1958, 1964, 1971).

While it is difficult to pinpoint the exact reason for such large and increased urbanisation among the female Chinese Australian population, between 1911 and 1966, the Chinese Australian population as a whole was becoming increasingly urbanised. Choi (1975) noted that rapid urbanisation of the Chinese Australian community after 1901 occurred as increased employment opportunities arose in capital cities and metropolitan areas. 'Urban' occupations such as furniture making, laundries, fruit and vegetable distribution, and market gardening in metropolitan areas replaced primary industries such as station hands and farm labouring as the primary employment among the (male) Chinese Australian population. At the same time, other economic opportunities, business and social connections and services offered by the Chinese Australian community increased in urban centres. Census data do not provide any evidence that can be used to explain why the female Chinese Australian population, in particular, became so strikingly urbanised (and why they were so concentrated in rural areas in 1911). However, we can assume that their shifts in settlement patterns were connected to these broader processes.

Again, due to the limitations of census data, first-hand information obtained from Chinese Australian women can be used to gain some insight into what specifically prompted their settlement patterns and the diversity of the settlement locations. All of the women interviewed for this project were residents in the greater Sydney region at the time of participation in the research. Their lives in the White Australia Period were, however, more geographically dispersed. The majority of the Australian-born participants (13 of the 19) were born in NSW—seven in Sydney and four in the regional towns of Warialda, Tumut, Glen Innes and Inverell. The remaining two Australian-born participants were born in regional QLD in the town of Texas and on Thursday Island. Therefore, a diversity of rural and urban locations is represented in the birthplaces of the Australian-born participants—a more diverse settlement in contrast to the census statistics. The Australian-born participants did not however always remain in their place of birth throughout the period. Of the seven participants who were born in Sydney, three did not remain in the city throughout the White Australia period, relocating to Cairns in Northern Queensland, Europe and Canada. The six participants who were not born in Sydney also experienced various degrees of mobility. Four moved directly to Sydney from their home towns while two moved to several rural towns in regional NSW and QLD before eventually settling in Sydney as young adults.

The movement experienced by the Australian-born participants was primarily dependant on family decisions or relocations to husbands' place of residence after marriage. For example, Eileen was born in Sydney but her family relocated to Cairns when she was one year of age in 1947. She did not return to Sydney until 1966 when she was 20 years of age. Her decision to return to Sydney was a result of her marriage to a Chinese Australian businessman whose produce company was based in Haymarket, Sydney. Similarly, Marina's move to Sydney from Inverell, a small town in northern NSW, occurred in her adult life at the age of 28 as a result of her marriage to a Sydney-based Chinese Australian. For Mabel, Sally and Edith, the move to Sydney from their town of birth occurred early on in their childhood. Mabel moved from regional NSW to the western suburbs of Sydney with her family at the age of five as her father had bought a farm and later opened a fruit shop. Similarly, Edith also moved from her town of birth (Glen Innes) in regional NSW to Sydney in her childhood. She recalled that when she was eight, her parents decided to move the family to Sydney, explaining:

> It was mainly the future of the children, I think, because they thought that my brother at that stage was just about to leave school and they thought that education would be better to be in the city and also I think my father was ready to move out of the business as well.

Therefore, like Mabel's experience, Edith's family's movement to Sydney was economically motivated. On the other hand, Sally's family was actually based in Sydney but she was born in Tumut (a regional town close to the nation's capital) in 1942. She wrote:

> As the war developed, mother, Annie, [and] Winnie [Sally's sisters] went to the town of Tumut due to fears for their personal safety. The CBD was at risk from aerial bombing from the Japanese. A blackout period was in place to

prevent the Japanese planes from easily spotting the CBD. It was during this period that [I] was born in Tumut.

(Written correspondence, Sally 2010)

At one year of age, Sally, her mother and sisters returned to Sydney.

The influence of family decisions on the mobility of Chinese Australian females was also evident in the lives of the two participants born in regional Queensland. Mei-Lin, born in the small town of Texas, QLD, moved with her family at the age of ten to Ballina in northern NSW. She explained:

> We went to Ballina because if we'd stayed in Texas my brother and I would have had to have gone to boarding school because it's such an isolated town. And then one of my uncles had a brother-in-law who had a general store in Ballina. He had four children and wanted to move them to Sydney for their education so he encouraged my father to buy this business thinking that Ballina was a bigger town with more opportunity for us children and Mum and Dad. In many ways it provided that, and meant that Mum and Dad had to work much harder in their later life […]. So Dad and Mum bought this business in Ballina and we moved across.

While Mei-Lin's parents decided to relocate for economic reasons and to give their children greater educational opportunities, Stella, who was born on Thursday Island, first moved to mainland Australia in 1943 during World War II, as she and her family were evacuated to Cairns. She recalled:

> Yes, we chose Cairns because it was the closest and she [Grandmother] wanted to high-tail it back to the Island as soon as the war broke out, thinking it would stop next year. […] We were four years in Cairns and as soon as we were allowed to, we took a little cargo boat back […] I only stayed there [back on Thursday Island] a year or two—about two years at the most—and decided to go back to go to my mother [in Mackay, Queensland] and I stayed with her a few months and took off to Sydney. I'd just turned seventeen.

Coincidentally, both Mei-Lin and Stella later experienced more 'independent' mobility compared to the other interview participants. At the completion of her high school education in Ballina, Mei-Lin separated from her family to attend Teacher's College in Armidale, another regional town in Northern NSW. She then moved to Sydney to take up her first teaching position. Stella, with the aid of the United Association of Women, was able to find accommodation and employment in Sydney, where she has continued to live. From these histories, it is clearly evident that the movement of the participants within Australia was based on a desire to improve opportunities—economic or educational. The ages of migration and specific reasons for migration were, however, diverse. For some women, movement was shaped by family decisions in their childhood, while for others, it was relocation due to marriage, while for Mei-Lin and Stella, their mobility in young adulthood was self-determined by their desire for further education and/or employment.

The migration stories of the foreign-born participants discussed in the previous section reveal international mobility; however, unlike their Australian-born counterparts, once in Australia, they generally experienced less movement around the country. Sydney provided the location of first settlement for the majority of foreign-born participants—four out of six settling in Sydney. All four of these participants—Daphne, Helen, Patricia and Lily—did not leave Sydney in the White Australia period. For these women, Sydney was a place in which they built their new lives—establishing or reconnecting family networks and finding education and employment. Mary, who arrived as a high school student, moved from Maitland in regional NSW where she attended boarding school to Sydney to finish her final years of high school at 'Sydney Tech' and did not relocate after. Sandy—the only migrant participant to first settle outside of NSW—made only one move from Melbourne to Sydney. Due to limitations of the interview data, it is not known why the immigrant participants did not leave Sydney once they had arrived. I am reluctant to suggest that their (non)citizenship or visa status effected their mobility within Australia as participants did not communicate a desire to relocate that could not be realised for that reason. I am also reluctant to assume that these women and their families had a timidness to move from the immigrant gateways. Given the accounts provided by the interview participants, it is perhaps more accurate to suggest that these participants remained in Sydney as it was a city that provided economic, educational and social opportunities and was a place in which family obligations or ties were located.

Conclusion

By utilising census data to highlight the presence of Chinese Australian women throughout the White Australia Policy era, I have demonstrated one way in which racialised information can be used for inclusive research. The presented reading of the census data indicates that the Chinese Australian population was complexly comprised of both 'full' and 'mixed', Australian-born and foreign-born Chinese Australian females, alongside their male counterparts. It also reveals that female Chinese Australian presence was enduring—dating back to at least the 1860s. This corrects longstanding assumptions that Chinese Australian females were absent in the White Australia Period and suggests alternative subject positions (i.e. positions beyond the 'absent wife') disregarded in the literature. For example, the large proportions of Chinese Australian females under 15 years of age indicate that Chinese Australian females were not only notable in the Australian community in their adult position as wives, as women in non-nuclear family arrangements, or as 'stubbornly' unmarried, but as daughters and children. As I have highlighted, 17 of the interview participants were included in this group, having spent their childhood and/or teenage years in Australia during the White Australia Policy era.

This chapter has also begun to explore Chinese Australian women's mobility throughout the period. By combining official birthplace statistics and migration data with qualitative accounts of migration to and from Australia (as told by Chinese Australian women themselves), we can not only see the presence of Chinese Australian women in this historical and geographical context, but also the diversity of movements of foreign and Australian-born participants. Women migrated as dependent children, students and wives, and did not always follow the traditional or expected route of the passive or dependant wife following a migrant husband. An analysis of the women's accounts of their own lives shows that some Chinese women occupied empowered positions in the migration process—being able to negotiate and work around immigration restrictions rather than being passive victims of discriminatory policy. Their motivations for this movement across international boundaries also varied—including study, economic opportunity and marriage—highlighting the diversity of the migration experience. The migration experiences recalled by the foreign-born interview participants, and those recalled by Australian-born participants of their female forebears, clearly indicate differences between early twentieth-century female migrations and those that occurred in the post-war period. Many of the migrant participants experienced mobility as children/students, while the mobility of the female forebears of the Australian-born participants was predominantly defined by their role as wives. Despite these differences, evidence of Chinese Australian females' international migration presented through these women's accounts further challenges assumptions of females' positions in patrilineal Chinese traditions and broader assumptions of female immobility in global migration patterns.

In the chapters that follow, I will draw upon and extend the discussions that I have introduced in this chapter. By further drawing upon information obtained from the 19 interviews and census data, I will provide more detailed and personal pictures of female Chinese Australian roles, contributions and interactions throughout the White Australia Policy era.

Notes

1 For example, after the Communist takeover and establishment of the People's Republic of China in 1949, the *Land Reform Act, 1950–3* not only eliminated large land ownership, but confiscated land owned by overseas Chinese families. These factors weakened family lineage systems and made the desirability of return migration minimal. Return migration itself was made difficult due to stricter immigration–emigration controls (Choi 1975: 57).

2 This chapter has been derived in part from Kamp, 2021, 'International migration and mobility experiences of Chinese Australian women in White Australia, 1901–1973', in K. Bagnall and J. Martínez (eds.), *Locating Chinese Women: Historical Mobility between China and Australia*, HKU Press, Hong Kong, 105–128 (Copyright Hong Kong University Press).

3 Daphne and Sally also married non-Chinese Australians but these unions occurred in the post-White Australia period.

4 In accordance with definitions of 'generation' used by the Australian Bureau of Statistics (2012), 'first generation Australians' are people living in Australia who were born overseas; 'second generation Australians' are Australian-born people living in

Australia, with at least one overseas-born parent; 'third-plus generation Australians' are Australian-born people whose parents were both born in Australia – one or more of their grandparents may have been born overseas or they may have several generations of ancestors born in Australia.

5 As Choi (1975: 62) argued, 'we have to be content with "nationality" figures, which, though undercounting large number of ethnic Chinese who have non-Chinese nationality, are probably slightly better than "birthplace" figures which have a high proportion of non-Chinese'.

6 It must be noted that Chinese females (including some interview participants) also migrated to Australia as young students. The insights that interviews with women who migrated as students are not, however, included here but are discussed in Chapter 4.

7 Except in 1911. Queensland in the pre-war years was also an exception with a relatively even spread of rural and urban females between 1921 and 1947. As Jones (2005) noted, Brisbane, the capital city of Queensland, had a relatively small population of Chinese Australians compared to other state capitals. Many of Queensland's Chinese Australian population resided in Cairns, Townsville, and other regional centres in the north of the state.

References

Australian Bureau of Census and Statistics (ABS). 2012, *Reflecting a Nation: Stories from the 2011 Census, 2012–2013*, Australian Bureau of Census and Statistics, Canberra.

Bagnall, K. 2006, *Golden Shadows on a White Land: An Exploration of the Lives of White Women Who Partnered Chinese Men and their Children in Southern Australia, 1855–1915*, PhD thesis, University of Sydney.

Bagnall, K. 2011, 'Rewriting the history of Chinese families in nineteenth-century Australia', *Australian Historical Studies*, vol. 42, pp. 62–77.

Choi, C. 1975, *Chinese Migration and Settlement in Australia*, Sydney University Press, Sydney.

Commonwealth Bureau of Census and Statistics (CBCS). 1917, 'Volume II-Part VIII. Non-European Races', *Census of the Commonwealth of Australia taken for the night between 2nd and 3rd April, 1911*, Melbourne.

Commonwealth Bureau of Census and Statistics (CBCS). 1924, 'Volume.1-Part V. Race', *Census of the Commonwealth of Australia taken for the night between 3rd and 4th April, 1921*, Melbourne.

Commonwealth Bureau of Census and Statistics (CBCS). 1937, 'Volume.1-Part XII. Race', *Census of the Commonwealth of Australia 30th June 1933*, Canberra.

Commonwealth Bureau of Census and Statistics (CBCS). 1951, 'Volume.1-Part XV. Race', *Census of the Commonwealth of Australia 30th June 1947*, Canberra.

Commonwealth Bureau of Census and Statistics (CBCS). 1958, 'Volume VIII. Australia. Supplement to Part I. Cross-classification of the characteristics of the population: Race', *Census of the Commonwealth of Australia 30th June 1954*, Canberra.

Commonwealth Bureau of Census and Statistics (CBCS). 1964, 'Census bulletin no. 36. Race of the population Australia, states and territories', *Census of the Commonwealth of Australia 30th June 1961*, Canberra.

Commonwealth Bureau of Census and Statistics (CBCS). 1971, 'Volume I. Population: Single characteristics. Part II. Race', *Census of the Commonwealth of Australia 30 June 1966*, Canberra.

Couchman, S. 2011, 'Making the "last Chinaman": Photography and Chinese as a 'vanishing' people in Australia's rural local histories', *Australian Historical Studies*, vol. 42, pp. 78–91.

Gittins, J. 1987, *A Stranger No More*, Published by author, South Yarra, VIC.

Inglis, C. 1972, 'Chinese in Australia', *International Migration Review*, vol. 6, pp. 266–281.

Ip, M. 1990, *Home Away from Home: Life Stories of Chinese Women in New Zealand*, New Women's Press, Auckland.

Ip, M. 1995, 'From gold mountain women to astronauts' wives: Challenges to New Zealand Chinese women', in P. Macgregor (ed.) *Histories of the Chinese in Australasia and the South Pacific*, Museum of Chinese Australian History, Melbourne, pp. 274–286.

Ip, M. 2002, 'Redefining Chinese female migration: From exclusion to transnationalism', in L. Fraser and K. Pickles (eds.) *Shifting Centres: Women and Migration in New Zealand History*, Otago University Press, Dunedin, pp. 149–165.

Jones, P. 2005, *Chinese-Australian Journeys: Records on Travel, Migration and Settlement, 1860–1975*, National Archives of Australia, Canberra.

Kamp, A. 2021, International migration and mobility experiences of Chinese Australian women in White Australia, 1901–1973, in K. Bagnall and J. Martínez (eds.) *Locating Chinese Women: Historical Mobility between China and Australia*, HKU Press, Hong Kong, pp. 105–128.

Klapdor, M., Coombs, M. and Bohm, C. 2009, *Australian Citizenship: A Chronoloy of Major Developments in Policy and Law*, Department of Parliamentary Services: Parliament of Australia, Canberra.

Ling, H. 1997, 'A history of Chinese female students in the United States, 1880s–1990s', *Journal of American Ethnic History*, vol. 16, pp. 81–109.

Ling, H. 2000. Family and Marriage of Late-Nineteenth and Early-Twentieth Century Chinese Immigrant Women. *Journal of American Ethnic History*, vol. 19, no. 2, pp. 43–63.

Loh, M. and Ramsey, C. (eds.) (1986), *Survival and Celebration: An Insight into the Lives of Chinese Immigrant Women, European Women Married to Chinese and their Female Children in Australia from 1856–1986*, Published by the editors, Melbourne.

Macgregor, P. 1998, 'Dreams of jade and gold: Chinese families in Australia's history', in A. Epstein (ed.) *Australian Family: Images and Essays*, Scribe Publications, Melbourne, pp. 25–36.

McKeown, A. 1999, 'Conceptualizing Chinese diasporas, 1842 to 1949', *The Journal of Asian Studies*, vol. 58, pp. 306–337.

Ryan, J. 2003, *Chinese Women and the Global Village*, University of Queensland Press, St Lucia, Queensland.

Tan, C. 2003, 'Living with 'difference': Growing up 'Chinese' in white Australia', *Journal of Australian Studies*, vol. 77, pp. 101–108, 195–197.

Williams, M. 1999, *Chinese Settlement in NSW: A Thematic History/A Report For the NSW Heritage Office*, NSW Heritage Office, Parramatta.

Wilton, J. 2004, *Golden Threads: The Chinese in Regional New South Wales*, New England Regional Art Museum, Armidale in association with Powerhouse Publishers, Sydney.

4 Domestic roles and the family economy

During my fieldwork, I often ventured into the homes of Chinese Australian women to hear and record recollections of their experiences during the White Australia Policy era. The centrality of 'home' not only figured in my research method—as a site in which participant and researcher could meet and share knowledge and experiences—but also as a site at the centre of investigation. As I discussed in Chapter 2, 'home' and household geographies are of increasing interest to postcolonial feminist geographers, particularly the ways in which the domestic space is experienced, constructed and utilised in diverse ways by different women. As such, rather than solely viewing home spaces as universal sites of patriarchal subordination, homes have been reconceptualised as complex and dynamic sites of oppression and marginalisation *as well as* safety, survival and emancipation. Homes have also been reconceptualised as sites, formations and processes constituted via the intersection of public and private spheres. That is, not only sites of the familial and domestic, but influenced by commerce, imperialism and politics (Blunt and Dowling 2006). In *Intersectional Lives*, such postcolonial feminist conceptualisations of home spaces operate as a framework to examine family dynamics, particularly the ways in which patriarchal structures within families and homes differed, or were experienced differently, by Chinese Australian females. I am also eager to understand how the public/private divides were blurred in homes (in relation to family dynamics and roles), or reinforced, and how these experiences played out across time/age and in relation to Chinese Australian females' diverse gendered, classed, generational, racialised and cultural subjectivities. Issues relating to 'home' and family therefore permeate, to varying extents, all the remaining chapters of this book. However, in this chapter, I stridently draw attention to the home by examining the domestic roles of Chinese Australian females and their contributions to family economies[1]. This begins with an examination of interview participants' childhood lives—the gendered nature of their upbringing and expected roles/contributions to the home and family. I then move on to explore female adult experiences as wives, mothers, adult daughters and contributors to the running of households and family economic stability.

This examination is imperative given the dearth of information available regarding female contributions to the endurance of Chinese Australian families and communities in White Australia. It is now well understood that within the

DOI: 10.4324/9781003131335-4

White Australia Policy context (male) Chinese immigrants and their descendants most often found employment in niche occupations that did not require a high level of language proficiency or officially recognised skills/qualifications (or indeed did not compete with 'Australian' jobs) such as employment in grocery, fruit and vegetable stores, general stores, market gardens, restaurants/cafes or furniture factories (May 1984; Rolls 1996; Fitzgerald 1997; Wilton 1998, 2004; Williams 1999; and Fitzgerald 2007). In fact, Australian immigration legislation exempted male Chinese from restrictions if they were providing non-competitive labour (Choi 1975: 41), and some Chinese Australian men were able to establish their own businesses in rural towns and large cities within these niche industries and provide sponsorship/employment to Chinese migrant men (whom they were often connected to by kinship, village or clan ties). The presence and contributions of women to these businesses and the economic survival (if not success) of Chinese Australian families and communities is largely overlooked or cast to the margins. Existing studies also tended to reinforce the dichotomy of the public/private spheres by focusing on male economic and political activities and relegating issues pertaining to home/family/women to separate chapters or sections (if included at all). This chapter offers a counterpoint by focusing on the roles and contributions of Chinese Australian women in homes, families and family economies during the White Australia Policy era. In doing so, it provides a more complex understanding of the interconnections between 'home'/family and economic survival.

Confucianism in Australian family contexts

When examining Chinese Australian women's experiences of home and economic participation in the White Australia Policy era, it is necessary to address Confucian family systems. The idealised Confucian family system, summarised in Chapter 1, has tended to be conflated by Western scholars with the reality of 'traditional' Chinese cultural practice. It is often assumed by Western scholars that patriarchal Confucian values are uniformly maintained by Chinese communities across historical, geographical, political and class boundaries (see Stacey 1976). Importantly, as Teng (1996) noted, Western scholarship has tended to assume that there is an essential Chinese culture—a 'traditional' culture rooted in Confucianism—that continued, unchanging over 'an unspecific historical period that is somehow understood to exist (sometimes coexist) in opposition to modernity' (Teng 1996: 134). Assumptions of 'traditional' Chinese culture and society have also been symptomatically Orientalist in their emphasis on cultural stasis or backwardness (Teng 1996: 134). These essentialist notions of Chinese culture fail to acknowledge regional, class or ethnic differences, as well as difference and change over time.

With the status of women often used as a measure of a society's development (Teng 1996: 121), Western feminists have been influential in the propagation of such Orientalist discourses, drawing on Confucian ideology as explained in the Classics[2] and constructions of women in May Fourth[3] discourses, as evidence of women's lowly status in 'traditional' Chinese culture. As such, the 'traditional'

Chinese woman has been constructed in relation to her victimised position within the oppressive patriarchal family system. While this oppression has been viewed as part of the universal subordination of womanhood, constructions of 'traditional' Chinese women have pointed to the particular oppression they experienced as a result of the extreme nature of Chinese patriarchy (Teng 1996: 124–5). Teng (1996) has highlighted that such assumptions have epitomised the phenomenon in Western feminist discourse of marginalising the 'Third World woman' and treating them as a 'singular monolithic subject' (Mohanty 1984: 333; see also Gandhi 1998).

In Chapter 3, I provided evidence that idealised Confucian gender roles did not consistently shape the experiences of Chinese Australian female migration and mobility. This follows the works of Wolf (1972) and Watson and Ebrey (1991) which have highlighted that the family structure proposed by Confucianism is not an ideal adhered to by all Chinese families across time and place. It also follows findings regarding the varied nature of the reproduction of patriarchy and gender roles in migrant Asian families (see, for example, Hugo 2002 and Lam et al. 2002). In this chapter, I will continue to challenge notions of Chinese cultural uniformity and the centrality of Confucian ideals. I will do so by demonstrating that Chinese Australian females' experiences in homes and families throughout the White Australia Policy era were not singular or even. Some women recalled experiences that reflected 'traditional' patriarchal/ Confucian values (such as having limited educational opportunities as children and later in life having responsibility for domestic work and childcare) but also often remembered roles that did not conform to 'traditional' gender expectations (such as work in 'the front'[4] of family businesses or alternative employment, and educational opportunities and achievement). In these ways, their experiences and roles were diverse and included various manifestations of participation, contribution and agency as well as patriarchy in the family and home. Furthermore, despite White, western assumptions that Chinese women are more oppressed than their Western counterparts due to the patriarchal systems aligned with Confucian ideology, I argue that interview participants' experiences of patriarchy (and that of their mothers) were not uniquely 'Chinese'. At the same time, they cannot be conflated with broader conceptions of Western patriarchy. I will indicate that their experiences were individually unique—shaped by patriarchal systems, cultural values and unique social and economic circumstances. The intersection of their ethnic/cultural identities, gender identities, class, age and generation also influenced their roles and contributions to homes and family economies in the White Australia Policy era.

Memories of childhood: creating future wives

For several participants who grew up in Australia, experiences within the home and family institution were shaped by the expectation that they would eventually and most certainly be 'married off', with their future roles as dutiful wives relegated to home and family duties inscribed from an early age. For example, in discussions of growing up on her father's market garden in a southern suburb of

Sydney in the 1940s and 1950s, Doreen (pictured in Figure 4.1 with her siblings) recalled the assumption that her future lay in her marriageability and the stark difference between the treatment of the male and female children in her family:

> My mother was very much the boys are what carry on the father's name and the girls just go and get married so we're not going to do anything. So whenever there was anything like new school shoes, or new school shirts the boys got it, but I had someone's hand-me-downs all along the line.

Figure 4.1 Doreen Cheong née Lee and her brothers, c.1947. Doreen and her brothers were all born in Sydney, Australia, after their mother emigrated from China to be reunited with her husband just before the outbreak of World War II. Doreen and her brothers spent the majority of their childhood growing up at their father's market garden in Kogarah, Sydney.

Source: Doreen Cheong (private collection).

Doreen went on to link these gender divisions to her educational opportunities: 'Back in the 1940s and 1950s the girls would just go off and get married so there was no point in them getting an education'. For Susan, whose Australian-born father ran a grocery store in western Sydney, memories of growing up in the 1940s and 1950s were similarly marked by unequal educational opportunities of sons and daughters:

> My father depended on us a lot. I remember when I was in Year 6 at school and my brother at that time was doing the equivalent of the HSC [Higher School Certificate]. I had to miss a lot of school that year because my father didn't want my brother to miss any school, but it was ok for me to so that I could help the shop and help him in the business. That was ok. It was just something that had to happen.

The inconsequential nature of Susan's education and her contribution to the family business was rooted in her gendered position within the home and family. This role enforced her training as a 'dutiful wife' by ensuring success was achieved via the male line. Implicit in Susan's recollection ('That was ok. It was something that had to happen') is the notion that such gendered expectations were natural, unquestionable and essential. It also, perhaps, refers to circumstances beyond the home and family, that is, social, political and economic circumstances which saw Chinese migrant families like Susan's encounter socio-economic barriers due to formal and informal racial discrimination in the White Australia context. Contributions of young daughters to the running of family businesses (and prioritisation of son's education) may have been essential to the economic survival and success of the family.

For most of the participants who experienced this type of childhood gender inequality, it was assumed that this was a typically or traditionally 'Chinese' way. For example, when I asked Doreen if she believed the difference in treatment of sons and daughters was a typically 'Chinese' way of thinking, she responded by saying:

> Oh yes, very much. [...] My father was a very soft natured man and all his children were equal but to my mother, she was the one that stressed how important it was for the children to pass on their father's name. And of course, back in those days, people had children because they wanted an insurance policy that their children would look after them in their old age and the ones that usually did that was the boys. So it was more important from that aspect because usually the girls married away from the family and again my mother, not having any role model, would see that once the girls were married that would be…we would be the boy's family's responsibility and so that I think coloured a lot of her thinking.

Doreen also directly linked her childhood experiences with Confucianism: 'There was a Confucianism that came through and Confucian thinking is that the girls don't matter. The emphasis with Confucian thinking is on the boys'. Doreen's recollections highlight that the Confucian emphasis on male lineage and the wellbeing of the family (over the individual) was indeed a priority for her mother and thus her childhood family—male members of her family took precedence in regards to material possessions but also in regards to educational achievement and success. Doreen therefore did not understand her gendered experiences as unique or individual, but as part of a shared collective norm among other Chinese in Australia and elsewhere. By drawing upon Orientalist assumptions of collective Chinese 'backwardness' to understand her experiences, Doreen normalised and defended her childhood family dynamics; it was part and parcel of their Chinese cultural identity.

It is interesting that Doreen remembered that it was her mother that enforced the strict patriarchal system in the family rather than her father; asserting that her mother did not have a 'role model' in Australia to teach her any differently. In this way, Doreen once again hints at a binary opposition between what she viewed as the archaic 'Chinese' and modern 'Australian' mentality. In fact, Doreen later made a more overt link between education and adherence to Chinese traditions in explaining her mother's superstitious ways:

I think the reason for that was that my mother wasn't educated. She was a poor little village girl who had been sent out to work herself when she was probably nine or ten in the rice paddy fields and so obviously she had no other role model to base what her values were. [...] Because of her lack of education she was much more superstition bound, much more stubborn.

When I asked Doreen if her father was more 'Western' thinking than her mother, she explained:

I think he was not so much Western but I think he let my mother carry on in her superstitious ways. I think he was just a more kind-hearted, soft-natured kind of person and all he wanted was his kids to do well at school. He realised that education was very important.

By explaining that the differences between her parents' views were not a consequence of one being more 'Western' than the other, but that her mother's more backward 'superstitious' mindset was associated with a poor rural upbringing, Doreen inferred that Confucian family values varied within China itself—across geographical locations as well as across class groups. Thus, despite the transplantation and adaptation of 'traditional' Chinese values from rural China to post-war Sydney, and Doreen's understanding of the 'universality' of 'Chinese ways', deeper analysis suggests that Doreen's perceptions of her parent's values challenge notions of the uniformity of Confucianism in Chinese society.

It must also be noted that in reality, such divisions of labour and space were not unique to Chinese ideology and practice. Western feminists have highlighted gender roles within the family unit in which women are relegated to the home/domestic sphere and charged with caring/nurturing duties such as cooking, cleaning and general running of the household, and men with the public sphere and associated economic and political roles (Greer 1970; and Finch 1996). Although most interview participants associated patriarchal gender roles and expectations with the 'traditional Chinese' or 'Confucian' mentality, it is important to acknowledge that the 'crossing' of patriarchal expectations across cultural boundaries was observed by some participants. Nancy (pictured in Figure 4.2 with her siblings), for example, explained:

My father felt that the boys needed better education because they'd be the breadwinners and everything. It was interesting; I thought it was just among the Chinese that that was the attitude. But actually, when I went to uni and we did our own backgrounds, it was interesting to see other students from Australian backgrounds, and they'd say the same things: the brothers were given more education opportunities.

Kaylin's reflections are also pertinent in this regard. When I asked her if she believed the differential treatment of boys and girls in her family was a particularly Chinese way, she explained: 'No, I think it was just general... a generational

Figure 4.2 Nancy Buggy née Loong with her siblings, Parramatta, Sydney, c.1945. Nancy (second from right) poses with her three brothers and sister. The family lived on top of Nancy's father's greengrocery in Parramatta until 1951. Nancy described having 'happy memories' of her childhood.

Source: Nancy Buggy (private collection).

thing. In those days the same thing happened, I think, in Western families. You know, the daughters were just expected to do all those things'. Thus, while other interview participants drew connections between patriarchal systems within their family and Chinese cultural norms, Kaylin believed that the different treatment she and her brothers received was not a particularly 'Chinese' phenomenon but rather, part of a broader historically situated social and cultural expectations.

The differing roles and expectations placed on Chinese Australian sons and daughters had real impacts on the educational achievement and opportunities made available to several of the interview participants. It was common for these participants to have completed their high school education at the Intermediate Level (the equivalent of Year 10) at around the age of 15 (instead of going on to matriculate in the equivalent of Year 12) before entering the workforce, gaining qualifications as secretaries or nurses, and/or getting married. In contrast, for those participants who had male siblings, their brothers went on to complete high school and gain a university education. This pattern of gender difference was clearly highlighted in my interview with Edith who finished school at the Intermediate Level and went on to become a secretary. When I asked Edith if her parents thought her education was important she explained: 'they thought education for the girls was important up to school, I think, probably not as important career wise, but to get through school I think was important to them'. When I then asked what her parents expected her to do once she finished school, Edith responded with:

> That's why I did a secretarial course. I wanted to study science but they thought I should be a secretary because it's much easier for a girl [...]. I could've done it after, but I didn't. But I don't think they would've been happy for my brothers to have...well certainly they wouldn't have become a secretary or something. My older brother had a trade but then the second brother went to uni. But I just don't think it was as important for the girls to go to uni as it was as the boys.

It is interesting that Edith explains that her decision to become a secretary was based on her parent's views of what was 'easier for a girl', not her actual ability: 'I could've done it after, but I didn't'. Therefore, while she admitted that her educational and occupational outcomes were based on patriarchal assumptions, she rejected the validity of those assumptions of women's and, more specifically, her own capabilities.

A complex negotiation

For Australian-born Kaylin, her parents' expectations of her education were a complex negotiation between patriarchal gender roles and the encouragement of female educational and economic successes. Kaylin explained:

> I think it must've been a difficult situation in the sense that he also felt that yes, daughters should stay at home and help the mothers to run the household. So on the one hand he wanted me to study, but he never minded if I went to work, as long as I had an education. Then I could stay home and help run the household. [...] they always expected me to be taking responsibility for things around the house. So I mean, my father wanted me to have an education, but I think my mother always wanted an offsider in the house.

While traditional gender roles were maintained within Kaylin's family structure, her father highly valued the education of all his children—whether male or female. Her parents' value of all their children's education was also reflected in their choice of schools—Kaylin attended a private girl's grammar school and explained: 'The alternative was a domestic science school, in those days. That would not have been the choice of my parents'. Although Kaylin explained that her parents expected her to 'stay at home' after completing her education, she later explained: 'My father always wanted me somehow to become a company secretary [for his furniture business]'. Such inconsistencies in Kaylin's account of her father's expectations can perhaps be seen as an indication of the complex negotiation of gender expectations and collective family interests within Chinese Australian families. Kaylin's, Doreen's and Edith's experiences also help indicate that within households in which traditional gender expectations and roles were central, the maintenance of those roles and experience of gender discrimination was not uniform or even.

While patriarchal gender roles and expectations did shape the childhood experiences of several participants, the interviews importantly indicated that gender inequalities associated with Confucian or broader patriarchal family systems were not experienced in the same way in the childhood lives of all interview participants. In fact, the majority of participants claimed that their childhood lives were not marked by gender discrimination or the expectation that they would become dutiful wives according to the Confucian ideal. For many, education was encouraged within the childhood home and was generally linked to parents' own life circumstances and eagerness for their children (including daughters) to be educated. For example, Mei-Lin (pictured in Figure 4.3 with her parents) specifically recalled that it was her parents' own limited education that fuelled their value in educating their children:

> They were very happy for me to have an education…my dad, because he didn't have one. [...] He was passionate about having a good education. They both were. But neither of them had very much education.

Figure 4.3 Mei-Lin Yum and her parents at the Prefects Induction, Ballina 1960. In 1954, Mei-Lin and her family moved to Ballina, a small town in northern NSW, where Mei-Lin's parents obtained ownership of their own store and the children had greater educational opportunities. Mei-Lin is pictured here, in her final year of high school in Ballina, with her parents.

Source: Mei-Lin Yum (private collection).

For Ina, it was her poorly educated widowed mother who encouraged her education:

> She [Mother] preferred us to study and she said, 'Never work in the shop, go and study, go and study'. And when we graduated and she came to the graduation ceremonies, I'm sure she was very proud, but it was all bewildering for her because she didn't really know what it meant. But at least we were educated.

Ina later linked her mother's encouragement of her educational achievement with the harsh realities of her mother's own life. She described the struggles her mother endured in Australia, being a widowed immigrant from 'a fairly poor background' trying to raise and provide for two children in the 1940s and 1950s. According to Ina, her mother believed education was a means through which her children could have a better life.

The encouragement of daughters' education was also inherently evident in life-histories of participants who migrated to Australia as students. While research is slowly emerging regarding the emigration of female Chinese students between the 1880s and more contemporary times, particularly to the United States, very little is known of the female Chinese students who arrived in Australia under exemptions from immigration restrictions[5]. First-hand accounts (beyond this project) have highlighted that females were part of a post-war privately funded student cohort which Huck (1968: 46) has argued comprised the majority of overseas Chinese students in that period. For example, Christine Ramsey has recalled her migration from Singapore to Australia in 1958 at the age of 17 to complete her matriculation at Adelaide High School and enter Adelaide University (Ramsey 1986). In her recollections, she described the movement as a usual part of her socio-economic and cultural context: 'It was the done thing for people of that socioeconomic group to send their children to overseas universities…To the Chinese, right across the board, scholastic achievement is of great importance' (p: 29). After her studies, Christine returned to Singapore and spent time in London and the United States with her Anglo-Australian husband. In 1967, Christine was able to re-enter Australia owing to her marital status in what she remembers as a context of 'strict immigration restrictions for Asians' (p: 32). Thus, experiencing migration as a student and a wife, Christine describes herself as 'a student who stayed' (p: 29). In a similar way, Moni Lai Storz moved to Melbourne from Malaysia in the 1960s as a young adult to attend university. Her recollections were documented by Giese (1997) and tell of her arrival in Melbourne in 1963, her experiences of identity negotiation at Monash University, and subsequent intercultural experiences.

With these exceptions (and Martínez's (2021) examination of two Australian-born university-educated women of Chinese heritage), investigations of student life (either Australian-born or migrant) have been male-focused or gender-neutral. The experiences of three interview participants that came to Australia as students therefore provide important insights[6]. Helen left Hong Kong in 1961 to join her brother and complete her high school education in Australia before going on to university study. While both her parents highly valued education for all their children—boys and girls—the decision for Helen to come to Australia was made by her mother:

> She [Mother] was the one who actually encouraged me at that time, not only to finish high school—she didn't finish high school herself—but go to university if I can because she said, […] 'you'll have nothing, because without education you wouldn't climb the ladder at all, your ability will be stagnated'.

Helen did successfully complete her high school education in Australia and went on to obtain two university degrees in Arts and Law before becoming a practising lawyer and member of parliament. Helen's educational and professional achievements, her parents' pro-education stance, and her mother's active role in the migration decision-making process are pertinent as they challenged patriarchal Confucian ideologies characteristic of 'traditional' Chinese family structures.

In a similar way, Sandy arrived from Hong Kong in 1962 to join her sister and complete her high school education in Melbourne. She described her position as a foreign student as a privilege:

> I came as a student so I was very lucky. […] I started my Leaving Certificate which is year 11. And then I did year 12, which is called Matriculation in those days, and then I went and studied at university. […] After that I went and did a Diploma of Mathematics at RMIT.

While Sandy and Helen arrived in Australia during the latter years of high school, Mary left Hong Kong in 1967 with her half-sister at age 14 to restart her high school education at a boarding school in Maitland, New South Wales. A point of difference in her history is that her (and her sister's) student migration was not based on her parents' value of their daughters' education. In Mary's terms, she was sent to Australia because 'I think everything was getting a little bit complicated at home'. In fact, Mary explained that her parent's had very 'traditional Chinese' views regarding education and dreams for their daughters:

> I don't think my father had any expectations from the girls. Basically they just expect us to eventually marry and wouldn't be his problem anymore. The sons definitely were expected to return [from overseas education] and help him, support him one way or another. But no, I don't think my father had any thought of the girls in terms of future, you know, whatever happens, happens sort of thing, with the girls.

Like interview participants who were born or grew up in Australia and experienced such gendered expectations within the childhood family, Mary explained that such 'old fashioned' views were a result of her father's lack of education and 'village' ways:

> My father grew up in a small village in China. [...] He never had a formal education to start off with, and he didn't travel very much. He did eventually come to Sydney for a couple of weeks or something. So I think he's very...his views are quite narrow and old fashioned.

However, Mary went on to suggest that this striking difference in gendered expectations within her childhood family actually had positive impacts on her and her sister's lives: 'In a way that sort of helped us. All the girls knew that we only had ourselves to depend on'. Her experiences therefore indicate a complex negotiation between 'traditional Chinese' values, encouragement of daughters' education, expectations of female subservience and realities of female agency within the family unit.

Like interview participants who felt their educational opportunities were limited due to their female position in the family, others who had quite opposite experiences also constructed an Orientalist binary between 'backward' Chinese culture and 'modern' Western/Australian culture. They often linked their educational opportunities and successes with, what they termed, the more 'Western' thinking households in which they grew up. For example, Mei-Lin acknowledged that her parents' passion for her education was a reflection of their rejection of 'traditional' Chinese culture. She explained that 'there was never any question' of her educational opportunities as '[t]hey were not of the ilk that girls should stay at home, and so as I say we really did have an Australian not a Chinese upbringing, really'. Margaret, who grew up in Sydney's western suburbs in the 1950s and 1960s with two other siblings, similarly explained:

> My parents weren't really typical Chinese parents. Some Chinese parents don't even like girls to go out and work or anything, to study. My parents encouraged us to get as much learning as we can and do what we like. Back then, we were given a lot of freedom to do whatever we want to do.

Although feminists such as Summers (1975), Spender (1980) and Greer (1970) have stridently argued the perpetuation of oppressive patriarchal systems in broader Australian society, in Mei-Lin and Margaret's views, the Australian context provided a means by which 'traditional' Chinese patriarchal systems could be rejected and more equitable life opportunities, particularly in regards to education, could be realised. In this way, they constructed a dichotomous relationship between the oppressive 'typical' Chinese culture and the more liberal Western or 'Australian' culture.

The first-hand recollections of educational opportunities (and their relationship to family dynamics) also provide important insights as census data on the education of Chinese Australian females are limited. The 1933 census report (CBCS 1937) is the only one of the period that provides specific information regarding the type of education or schooling of those present in Australia (according to 'Race' and gender). The report indicates that of the 3,137 Chinese Australian females ('Full' and 'Half Caste') present in Australia on census night of 1933, 625 were receiving an education— 489 (78 per cent) in government schools, 125 (20 per cent) in private schools, and the remaining minority at university (two females) and home-schooled (nine females). If we assume the age of schooling to be between 5 and 19 years of age, we can estimate that of the 994 Chinese Australian females in this age group, approximately 63 per cent were receiving an education. This compares to 62 per cent of 'European' females in that age group receiving an education. While this can only be taken as a rough estimate as it cannot be known from the census data the exact ages of females in the education system (particularly in regards to University education), the statistics further reiterate the need to reconsider assumptions of the universality of Confucian ideals and practises across the Chinese diaspora and the disparity between 'Western' and Chinese female roles. The statistics also point to the need for further investigation.

Homemaking and contributions to the family economy

In line with existing literature on Chinese Australian economic activity in the White Australia Policy era, interviews with Chinese Australian women indicated that fathers usually played a central role in ensuring the economic stability of the family. They were primary breadwinners and often entrepreneurs, having established their own businesses such as grocery, fruit and vegetable stores, general stores, market gardens, restaurants/cafes or furniture factories in rural towns and larger capital cities. Unlike existing literature, however, interviews with Chinese Australian women also shined light on the hard work mothers/wives contributed to the running of the household. For example, Patricia explained that her mother did not participate in paid employment after their migration from southern China to inner-city Sydney, but instead 'stayed at home, looked after the family, cooked and all that'. In a similar way, Australian-born Nancy and Kaylin, who also grew up in Sydney recalled: 'Mum did the cooking, Dad worked'(Nancy) and '[Mum] never went to work. […] Well she had five children to look after'(Kaylin). Some participants also highlighted that once in the position of wife/mother themselves, they similarly had sole responsibility for household duties and the raising of children. For example, Marina, Eileen, Susan and Patricia recalled their primary role in their homes in Sydney while their husbands ensured the financial security of their families. All four women had participated in paid work in their young adult lives; however, ceased employment after their marriages. Reasons for ending their employment included assertions that their husband 'earned enough' (Marina) or, in Eileen and Patricia's cases, they bore children and subsequently 'stayed at home' (Eileen).

The experiences of Patricia, Nancy and Kaylin's mothers, as well as Marina, Eileen, Susan and Patricia themselves, seem to reflect an adherence to Confucian (and Western) gender expectations. However, it must be noted that none of these participants explicitly framed these gender roles as particularly 'Chinese' (or 'Western'). Similarly, despite the gendered allocation of family responsibility, participants did not necessarily view their domestic roles with antipathy. In fact, Patricia viewed her household and family contributions in a positive light. She expressed gratitude for her experience raising her children: 'I enjoyed staying at home with my kids. I wanted to see them grow up. So I was lucky in that way, I could stay home'. In this way, Patricia's role in the family can be viewed as an important contribution (rather than a form of patriarchal oppression and confinement). This challenges dominant Western feminist (particularly socialist and radical feminist) assertions that women's position in the family/domestic sphere is equated with female oppression (Summers 1975; Delphy and Leonard 1992; Greer 1999; and Finch 1996) and confirms postcolonial feminist assertions that 'home' and household are constructed differently for and by different women (Silvey 2006; and Blunt and Rose 1994).

Pushing the boundaries of 'women's work'

It must once again be acknowledged that Chinese Australian women's experiences in homes and families throughout the White Australia Policy era were not singular or even, as experiences did not always conform to clear gendered divisions. Susan, for example, found that as the constitution of her adult family changed with the death of her husband, she had to take on the 'double burden' of being the sole provider and primary carer of her children—roles that transgressed the traditional (Confucian and 'Western') gender divisions of space and labour and were necessary for the survival and success of her family. Other women who were wives in more traditionally 'nuclear' family units also had roles and responsibilities that extended beyond the domestic realm. Edith, for example, helped her husband in his business in Sydney while at the same time she was '[…] running the home and looking after the kids mostly and that sort of thing'. Doreen similarly helped her husband run his jewellery business in Sydney: '[My husband] ran a jewellery business and I would have to do the book keeping and send off the monthly tax returns and do the valuation'. Doreen also attended university, but at the same time, her husband enforced strict patriarchal restrictions:

> When I said I was going to uni, [my husband] said he didn't mind me going to uni provided nothing changed. So as long as all the housework was done and the children were looked after and meals were cooked, that was quite okay, he would let me go to uni.

The uneven power balance in Doreen's relationship with her husband is striking. Her reflections insinuate that she had to gain permission from her husband to attend university: 'he would let me go', and was delegated household/family responsibilities by him. Thus, even though Doreen participated in the 'public' sphere of university and employment, her husband asserted his patriarchal authority by defining the type of work Doreen had to conduct within the home/family unit and how she spent her time. These experiences indicate complex negotiations or flexibility between 'inside' and 'outside' work, or the 'public' and the 'private' spheres, within the confines of gender expectations.

Decades earlier, in some participants' childhood lives, participants' mothers also took on a dual responsibility—taking care of domestic duties as well as contributing to the running of family businesses. For example, Ina, whose father ran a fruit and vegetable shop on Sydney's Parramatta Road in the 1940s, remembered that her mother was responsible for the care of the children as well as cooking for the shop workers:

> [Mother's] function was out the back looking after children. When my father was alive he hired people to buy and sell and then went down and bought the stuff in the market and brought it back in a truck and then they'd unload it. He'd have about six people actually, and in those days the people lived on the premises and my mother would cook these huge meals and there'd be a round table and we'd all sit down and eat them. And you know breakfast was a full meal with rice and everything; it wasn't just a piece of bread. And lunch was fully cooked and she had a busy time just keeping the food up.

While Ina's mother's double responsibility for running the household and contributing to the family business transgressed gendered norms in some regards, her specific role in the family business—to provide meals for staff—can be viewed as an appropriation of the gendered expectation that women are responsible to 'feed the family' in their caring role (DeVault 1991). Furthermore, Ina's recollections of the daily routines of her parents indicate a gendered division of labour that placed her mother and father within very different physical spaces. Just like many British Chinese women working in family-run takeaway stores (Song 1995), Ina's mother was 'out the back', removed from the more public space of the shop and interactions with customers while her father was mobile and ventured into the public economic sphere of the market and managed the shop and its staff. It is also interesting to note that Ina's mother was illiterate in both English and Chinese and indeed never learnt to speak the English language. In contrast, Ina's Australian-born father was schooled in both Australia and China in the first two decades of the twentieth century. The language and educational differences between Ina's parents, itself an outcome of gender and class unevenness, was perhaps an underlying factor in this division of labour and space (as also found by Song 1995 in the British context). Despite such divisions, Ina's mother's contributions to the family business pertinently illustrate the interdependency of the public/business and private/domestic spheres in the constitution of family economies within Chinese Australian families in this historical context.

Ina normalised this gendered division of space and labour by describing her mother's role as a 'function'. This normalisation was further communicated in Ina's evaluation of her mother's experiences:

> And the interesting thing was, she never felt hard done by or isolated or whatever a woman of today would feel. She just got on with her job of washing—often by hand, by choice—and cooking and cleaning in fairly primitive conditions at the back of the fruit shop.

Implicit in Ina's appraisal is that her mother was indeed 'hard done by' and 'isolated' as a result of her work in 'primitive conditions'. Despite depicting her mother's experiences in terms of oppression, Ina asserted that her mother did not complain as it was 'her job' and duty as wife and mother to take on this double responsibility to the home and family business—thus echoing patriarchal assertions of women's 'natural' and 'instinctive' propensity to undertake menial domestic and/or unskilled tasks (Summers 1975: 175). Interestingly, Ina also alluded to elements of choice and control on behalf of her mother and thus, despite the conditions in which her mother found herself, attributed her with some power. In this way, Ina's memory of her mother's contributions to the running of the household and family business was coloured by a sense of pride and, once again, indicates the varied ways in which 'homemaking' and the domestic sphere are constructed/experienced by different women[7].

Challenging gender expectations

While Ina's mother contributed to the family business 'behind the scenes' by working 'out the back', for other mothers the double responsibility to the home and family economy saw them participating in more visible roles in 'the front' of businesses. For example, the prominent position of Chinese Australian wives in the running of family businesses was highlighted in Eileen's interview. Her parents (Australian-born Frederick Lee Yan and Chinese-born Fong See Lee Yan née Wai Hing) were particularly active in terms of their entrepreneurial exploits—first opening a noodle shop in Cairns' early Chinatown in the early 1940s (before Eileen was born), then a grocery/corner store, and eventually a fish and chips shop. All the while, Eileen's father, Frederick, took on extra work in a range of occupations. Eileen explained that because of her father's additional work, it was her mother who had prime responsibility for the family businesses, despite her inability to read or write English. When I asked if her parents worked alongside one another, she replied: 'No, because Dad…most of the time he had another job and Mum was taking care of the shop.' In Figure 4.4 Eileen's mother, Fong, stands outside her and her husband's first business—the noodle shop in Cairns.

Eileen explained that her mother's pivotal role in the family businesses was essential to the economic well-being of her family. When I asked if her mother's responsibilities challenged the idea of a 'traditional' Chinese wife whose role it was to stay home and look after the children, Eileen responded with: 'You had to, to make ends meet.' In a similar way, when I asked Susan, whose childhood home was above her father's fruit shop in Parramatta, if her mother had a particular role in the family, she explained: 'Well, she was always the mother. She had to help in the fruit shop, of course'. Both Eileen and Susan's responses, as declarative statements of fact, perhaps refer to the particular socio-economic barriers facing Chinese immigrants in the White Australia Policy era and indicate that their mother's dual responsibilities to the home and family business were subsequently unquestionable and obvious—their mothers *had* to help in the business, as a matter of course. Thus, despite taking on 'non-traditional' roles—both in terms of Chinese Confucian ideals and Western norms—their contributions were seen as natural and normal within the social, political and economic context of the time.

Figure 4.4 Fong See Lee Yan and first-born daughter outside her noodle shop in Cairns, Queensland, in the early 1940s.

Source: Eileen Yip (private collection).

While Ina, Eileen and Susan only implicitly suggested that the dual responsibilities of their mothers were 'non-traditional', Sally made explicit reference to the 'non-traditional' experience of her mother. Her father, Hung Jarm Pang, opened the famous Modern China Café in Sydney's Chinatown in 1940 and Sally explained that her mother, Choon Ping In (pictured in Figure 4.5), played a central role in the business: 'Traditionally she would be at home, but no, she was working on side with my father, more like handling the money...the cash register'. Sally's mother not only helped in the running of the business within the 'front' space of the Café but contributed on seemingly equal footing as her husband. Interestingly, Sally drew upon the idea of an intrinsic and immutable 'traditional' Chinese culture (which her mother rejected) to emphasise the agency and contribution of her mother to the economic survival and success of the family.

Women's contributions to family businesses were not confined to their position as a wife/mother. For Kaylin and Marina, contributions to the running of the family business were most prominent in their young adult lives, as daughters within entrepreneurial families. Kaylin did work as the assistant company secretary of her father's Sydney furniture business, while Marina worked at her father's department store from the age of 15. Due to the success of Marina's father's Hong Yuen store in Inverell and branch stores across rural NSW, Marina has often been asked by researchers to recall her father's life and his business enterprise. Emphasis is generally placed on the specific business practices and successes of her father, Harry Fay (Louie Mew Fay), while Marina's own experiences working in the store are largely overlooked (see, for example, Wilton 1998). When asked about her own experiences working at Hong Yuen, Marina remembered her interactions with customers and her eventual promotion to the position of manager of the women's department: 'I became the manager of that section. My sister got married, so I had to do the buying—come down to Sydney and Melbourne and do the buying. It was a lovely department. I loved it'. In addition to these positive recollections of her everyday work, Marina also provided photographs of the Hong Yuen women's department (Figure 4.6). The image, alongside Marina's verbal recollections of dealing with customers, managing staff, travelling and buying merchandise, provide important snapshots of her everyday working life at Hong Yuen. Together, her memories and photographs provide evidence of the contributions she made to the family economy and the ways in which such work within the public sphere or 'front' of the family business transgressed patriarchal divisions of labour and space.

Figure 4.5 Sally Pang with family members, 1969. Sally (back left) poses with her parents (front) and sisters Winnie (back centre) and Anne (back right) at Winnie's nursing graduation at Royal Prince Alfred Hospital, Sydney.

Source: Sally Pang (private collection).

The unquestioned and necessary contributions of wives and daughters to the running of entrepreneurial businesses can be viewed within the broader family business strategy employed by Chinese Australians throughout the period. Research has stressed the importance of family and kinship links in the establishment and running of 'Chinese' businesses in Australia (see, for example, Wilton 1998, 2004). Large businesses were often set up or expanded via a pooling of resources between several business partners who were often (but not exclusively) related by family, clan or village networks (Wilton 1998). Once established, family members were often employed and/or clan and village networks were drawn upon to provide labour (as was the case in Marina's and Kaylin's young adult lives). Often potential workers in China were sponsored provided that the sponsoring

Figure 4.6 Staff of the Hong Yuen lady's showroom, Inverell, 1949. Marina Fay (second from right) spent her childhood with her seven siblings in Inverell and began working at Hong Yuen's at the age of 15. She eventually became the manager of the lady's department.

Source: Marina Mar (private collection).

business was able to financially support the sponsored migrant. It was through this proviso in the legislation that merchants and other entrepreneurs were able to sponsor family members and ensure their successful entry into the nation (Wilton 1998: 99). Within the context of discriminatory legislation, it was in these ways that business partners and employees sought to support and work with family members and people from the same village and district in China (Wilton 2004). The 'traditional' Chinese emphasis on the centrality of family and kinship was thus translated into the business arena. The majority of businesses established by interview participants' fathers were, however, small business enterprises which did not require multiple shareholders or sponsored employees. Nonetheless, as I have illustrated, the contributions of the family as a labour supply were similarly paramount within the social, economic and political context of the White Australia Policy era (as similarly argued by Li (1993) in regards to the Canadian context and Song (1995, 1997) in the British context). While the literature has emphasised the important contributions of men in these roles throughout the White Australia period, women's unique recollections of their and their mother's everyday lives, highlight that they too were crucial contributors to the success of family businesses in ways which blurred or challenged gendered public/private dichotomies.

Employment beyond family businesses

In her examination of Chinese Australian settlement and contributions in regional NSW, Janice Wilton (2004) asserted that Chinese Australian (migrant) women who sought paid employment outside family businesses were rare. She explained:

> Given the drastic gender imbalance among the Chinese population—traditionally only Chinese men emigrated—there were few Chinese women seeking paid work. The small number of women who were born in China and migrated to Australia came primarily as wives or daughters and managed the household domestic duties and assisted in family businesses.
>
> (p.43)

Like Wilton, my research indicated that it was indeed rare (at least amongst the participant cohort) for migrant mothers to participate in paid employment beyond roles in family businesses. This rarity is reiterated in the census data for the years 1911, 1921 and 1933 (the only census years to record occupational characteristics of the population by 'race'; see CBCS 1917, 1924, 1937). However, it must be acknowledged that while the paid employment of Chinese Australian females was rare, it was not unheard of. For example, in 1911, 2,053 Chinese Australian females indicated that they were dependents. The majority of these dependents (594 or 25 per cent of all Chinese Australian females) were wives/mothers/widows who undertook domestic duties for which remuneration was not paid. Daughters and relatives who were attending school or being taught at home (578 or 24 per cent of all Chinese Australian females) were also numerous, as were dependent daughters and relatives who were not stated to be at school or performing domestic duties (586 or 24 per cent) (see Table 4.1). Despite this high proportion of dependency, 14 per cent (n: 342) of Chinese Australian females were in paid employment (see Table 4.2). In 1921, the rate of employment among Chinese Australian females increased to 16 per cent (480 of 2,924 women in 1921) and in 1933 to 17 per cent (543 of 3,137 women in 1933) (see Table 4.3). The 1933 Census Report also indicates that 773 female Chinese Australians were classified as 'breadwinners' of their family/household. Within this breadwinning group, 111 did not make any income for the year ending 30th June 1933, while the majority (n: 287) earned £287 and 20 females fell into the highest income bracket, earning over £260 that year. Although the majority of women in 1911, 1921 and 1933 were not formally recorded as being employed, the census data do indicate that paid employment among the Chinese Australian female population increased over the three decades and that some Chinese Australian women did contribute to the Australian economy. Most striking are the data on female 'breadwinners'— women who contributed the majority of their family income. Although small in number, the presence and contributions of these women challenge existing notions of the subordinate and dependent woman confined to the domestic sphere in accordance with patriarchal ideals.

Table 4.1 Chinese Australian Female Dependants in Australia, 1911[1]

Dependant Status	'Full Chinese'	'Mixed Chinese'	Total
Domestic duties for which remuneration is not paid			
Wife, mother, widow	279	315	594
Son, daughter, relative	57	219	276
Visitor	1	1	2
Boarder, lodger	2	9	11
Dependant scholars and students			
Son, daughter, relative and others at school	223	351	574
Son, daughter and others taught at home	1	3	4
Dependant relatives and others not stated to be performing domestic duties			
Son, daughter, relative (including persons under 20 years of age with unspecified occupations	267	319	586
Criminal class (under legal detention)			
Inmate of reformatory, industrial school	1	–	1
Inmate of a benevolent institution	–	1	1
Inmate in hospital for insane	–	3	3
State child	–	1	1
Not Stated	1	1	1
Total dependants	831	1,222	2,053

1 Figures for 'Full Chinese' have been appropriated from the category 'Full Blood Chinese' and 'Mixed Chinese' from the category 'Half-Caste Chinese' as defined by the Commonwealth (later Australian) Bureau of Statistics.

Source: CBCS (1917).

The participation of Chinese Australian women in paid employment is most pertinently reflected in the Census Report of 1911. It provides the most detailed account of the type of work conducted by Chinese Australian females and in doing so provides insight into the day-to-day activities of Chinese Australian females at the time. According to the report, of the 342 employed females, 16 (5 per cent of employed) were professionals—including one officer of a government department, one 'Clergyman/Priest' [sic], one architect and seven employed in the area of music (see Table 4.2). The majority of the women (n: 170 or 50 per cent of employed) were employed in domestic work—including employment in hotels and restaurants, laundries and private houses. Some of the women (n: 48 or 14 per cent of employed) were employed in commercial enterprise—including one butcher/meat 'salesman' [sic], ten greengrocers and 22 shopkeepers. Industry, however, was the second most popular employment category with 87 women (25 per cent) employed in this sector. This included 77 women employed in clothing manufacturing, five milliners/stay-makers/glovemakers, and one metal worker. These data provide a picture of female contribution in the wider public domain during the early years of the White Australia Policy era.

Table 4.2 Occupations of Chinese Australian Females, 1911[1]

Occupation		'Full Chinese'	'Mixed Chinese'	Total
Professionals	Officer of Government Department	1	–	1
	Clergyman/Priest	–	1	1
	Hospital or asylum officer or attendant	–	3	3
	Midwife, accoucheuse, monthly nurse	–	2	2
	Architect	–	1	1
	Schoolmaster/schoolmistress/ teacher	–	1	1
	Professional Musician/Vocalist/ Student of Music	2	–	2
	Music Professor, teacher, etc.	2	3	5
Domestics	Hotel keeper, manager, servant	–	18	18
	Coffee Palace, restaurant, tea room, eating house, servant	1	4	5
	Board, lodging-house keeper, servant	–	5	5
	House servant	16	116	132
	Personal attendant	–	1	1
	Domestic nurse	2	–	2
	Mangler, laundrykeeper, laundryman, washerwoman	1	6	7
Commercial	Fancy goods dealer	1	–	1
	Shoe/boot dealer	–	1	1
	Butcher, meat salesman	–	1	1
	Greengrocer, fruiterer, potato, onion dealer	2	8	10
	Tea merchant/dealer	1	–	1
	Grocer	3	4	7
	Storekeeper, shopkeeper	9	13	22
	Clerk, cashier, accountant	–	1	1
	Commercial traveller, canvasser, salesman	1	2	3
	Telephone officer	1	–	1
Industrial	Machinist, stereotyper and others engaged in printing	–	2	2
	Lithographer, lithographic, zincographic printer	–	1	1
	Paper bag, box maker	1	2	3
	Clothing manufacturer, tailor, dressmaker	16	61	77
	Hat, cap, bonnet maker	1	2	3
	Shirtmaker	–	2	2
	Milliner, staymaker, glovemaker	1	4	5
	Bootmaker, shoemaker	–	1	1
	Other in industrial dress	–	3	3
	Baker, biscuit, pastry maker	–	2	2
	Fruit preserver, jam, pickle, sauce maker	–	1	1

(Continued)

Table 4.2 (Continued)

Occupation		'Full Chinese'	'Mixed Chinese'	Total
	Cordial, aerated water manufacturer	–	1	1
	Tea blender, taster, packer	–	1	1
	Tobacco, cigar, cigarette manufacturer	–	3	3
	Working in metals not elsewhere classed	1	–	1
	Factory manager, worker	–	1	1
Primary Producers	Marker gardener	1	–	1
Total employed		64	278	342

1 Figures for 'Full Chinese' have been appropriated from the category 'Full Blood Chinese' and 'Mixed Chinese' from the category 'Half-Caste Chinese' as defined by the Commonwealth (later Australian) Bureau of Statistics.

Source: CBCS (1917).

Table 4.3 Employment of Chinese Australian Females, 1921 and 1933

Grade of Occupation	1921			1933		
	Full Chinese	Mixed Chinese	Total	Full Chinese	Mixed Chinese	Total
Employer	6	11	17	21	15	36
On own account	22	53	75	42	46	88
Wages or salary	105	283	388	168	222	390
Apprenticed wage earner[1]	n/a	n/a	n/a	2	3	5
Wage or salary earner employed part-time[1]	n/a	n/a	n/a	9	15	24
Assisting without wages	5	9	14	14	6	20
Unemployed	7	21	28	44	62	106
Not applicable	997	1,394	1,191	1,235	1,231	2,466
Not stated	4	7	11	–	2	2
Total	1,146	1,778	2,924	1,535	1,602	3,137

1 This category was not included in the 1921 census. Persons who would have identified with this category were most likely documented under the category of 'wages or salary'.

Source: CBCS (1924, 1937).

Following her assertion that migrant women rarely sought paid employment, Wilton went on to explain that: 'Australian-born daughters increasingly joined the paid workforce, at first primarily in family businesses, paid and unpaid, and occasionally outside as, for example, dressmakers' (Wilton 2004: 43). Generational differences cannot be easily tested by the limited census data, but Wilton's argument is reflected in the lives of the women who took part in this study. Unlike migrant mothers (and fathers) who participated in paid/unpaid work in niche occupations that did not require a high level of language proficiency or officially recognised skills/qualifications (or indeed did not compete with 'Australian' jobs), Australian-born/Australian educated interview participants found employment in a diverse range of occupations that paralleled their educational achievements and qualifications. For example, six participants were employed in education (from early childhood teaching to high school teaching) and two were nurses/midwives. Participants were also employed as secretaries in schools, legal firms and private companies, as chartered accountants, news typists, scientists and social workers. In fact, all of the Australian-born women took part in paid employment at some stage of their life. This was clearly in opposition to Confucian ideology and gender expectations. The experiences also support Inglis' (1972: 270) and Bagnall's (2004: 160) assertions that education in Australia equipped Chinese Australians with skills, 'Australian' social and cultural knowledge, and recognised qualifications that allowed their full participation in White Australian society—including employment.

Conclusion

Through a postcolonial feminist approach, I have illustrated the complex and diverse nature of Chinese Australian females' roles and contributions to homes and family economies in the White Australia Policy era. From the outside, what may have seemed like traditionally patriarchal family households/systems in which female children and adults were allocated subservient and dependent positions was, in fact, much more complex. For example, the educational opportunities of Chinese Australian daughters and the way in which individual family dynamics and structures impacted these young women's lives were varied and diverse. Mothers and fathers were mixed in their expectations of daughters, with neither parental group being uniformly 'traditional' or 'non-traditional' in their outlook. With neither mothers nor fathers being more or less inclined to hold traditional patriarchal views of daughters' roles and responsibilities to the family, and given the differences evident within and across families, it may be logical to assume that the parents' birthplace (overseas or in Australia) influenced their perspectives. Perhaps migrants were more inclined to hold and maintain traditional cultural values. However, information obtained from interview participants suggests that this was not the case. According to the participants' recollections, Australia-born parents exhibited both 'traditional' and 'non-traditional' views (see, for example, Nancy vs. Mei-Lin), as did migrant parents (e.g., Doreen vs. Ina), and even parents who remained in China or Hong Kong and sent their daughters to study in Australia (Mary vs. Helen). While participants such as Doreen and Mary connected low socio-economic positioning and lack of education with 'backward' patriarchal gender expectations (and thus reiterated Orientalist views), recollections of other interview participants such as Ina and Mei-Lin indicate that there was no consistent link between the maintenance of patriarchal gender expectations within the home and class and socio-economic position. Unlike Doreen and Mary, both Ina and Mei-Lin asserted that it was their parents' own lack of opportunity (rooted in their low socio-economic position) which motivated their desire for their daughter's success beyond the strictures of the 'traditional' Chinese family system. With the majority of participants growing up in Australia in the 1940s and 1950s (albeit with some participants having experienced childhood in the 1920s, 1930s and up until the 1960s), differences among participants' gendered upbringing cannot be temporally accounted for.

Later in life, as wives and mothers, the diversity of experiences was similarly seen. Some women contributed to the running of households in ways which conformed to traditional gender expectations—cooking, cleaning and caring for children. Some also helped in the running of family businesses—either 'out the back' in ways that conformed to their traditional gender roles, or in more visible positions where they took on greater responsibility than traditional gender roles would prescribe. Others took up employment outside of the family business. This blurring of the boundaries between traditional gender expectations and the public and private via the uptake of the 'double burden' of maintaining the home as well as working in family or other businesses was a reflection of the specific social context. With discriminatory policies as the backdrop, these women were part of ethnic minority/migrant families that were trying to establish businesses and secure the financial stability and success of the family. As such, what was 'natural' or inherent according to traditional Chinese gender roles or Confucian ideals (as well as Western discourses) was not always possible for Chinese Australian families. Women could not always be an invisible presence in homes, but were often visible in family-run businesses by adapting 'female' roles to the business environment or in some cases challenging patriarchal divisions of labour and responsibility. In this way, the 'home', as defined by many Western feminists, was not simply a site of female oppression but rather a geographical location in which the public and private realms often intersected via women's contributions to family economic activities and domestic duties. 'Homemaking' was subsequently an act of negotiating, adapting and challenging patriarchal systems of power and crucial to the success of Chinese Australian families within and beyond the White Australia Policy era.

These findings are important as they highlight the multifarious ways in which gendered and cultural identities, and the power structures associated with them, impacted Chinese Australian females' experiences in homes and families. The diversity in experience indicates that there was no single 'traditional Chinese family' in Australia, which consistently dictated females' opportunities, responsibilities and contributions. Thus, it provides evidence that Confucianism is not an ideal adhered to by all families of Chinese cultural background and across historical and geographical contexts. The experiences of Chinese Australian females were not only influenced by familial attitudes and structures, but also by the broader context of White Australia which saw Chinese immigrants and their descendants marginalised in all spheres of life. In this way, 'the family' was not simply an institution of female oppression, but a social unit in which patriarchal systems of power were negotiated, adapted and challenged.

Notes

1 This chapter is derived in part from Kamp 2018, 'Chinese Australian women's 'home-making' and contributions to the family economy in White Australia', *Australian Geographer*, vol. 49, no. 1, pp. 149–165, copyright the Geographical Society of NSW, available online: http://www.tandonline.com/DOI:10.1080/00049182.2017.1327783. This chapter is also derived in part from Kamp 2021, 'Chinese Australian Daughters' Experiences of Educational Opportunity in 1930s–60s Australia', *Australian Historical Studies*, pp. 1–18 copyright of Editorial Board, *Australian Historical Studies*, available online: DOI 10.1080/1031461X.20201868543.

2 The 'Classics' refers to the Four Books and Five Classics of the 'Confucian' canon.

3 'May Fourth' refers to the student demonstrations that occurred on 4th May 1919 in protest against China's signing of the Treaty of Versailles. It also refers to the larger movement which ensued that was premised on a rejection of Chinese imperialism and tradition, and thus represented 'an entirely new type of grass-roots politics based largely on nationalist feelings' (Zarrow 2005, p: 149). In this political and cultural context, May Fourth reformers became increasingly influenced by Western notions of women's status and emphasised the victimisation and oppression of women in 'traditional' China. In fact, during the May Fourth Era (1917–1924), 'woman' became the symbol of the struggle between tradition and modernity (Teng 1996, p: 117).

4 Song (1995) adapted Goffman's (1959) conceptualisations of 'the front' and 'the back' (although unacknowledged) in her examination of the gender divides in the running of Chinese take-away businesses in Britain. I have similarly adapted these terms and used them in reference to other family businesses in Australia. In this way, I refer to 'the front' as the retail/serving space of restaurants, cafés, grocery stores, fruit and vegetable stores, etc., in which individuals had direct contact with customers, while 'the back' refers to areas of businesses not in public view such as kitchens and sorting/packing areas.

5 See Choi (1975) for an outline of changes to student entry provisions, Huck's (1968) investigation of Chinese students in Australia and Kuo and Fitzgerald (2016) and Martínez (2021). See Ling (1997) for the US context.

6 I will examine Chinese Australian female's experiences in school and university settings in Chapter 6: 'Interactions'.

7 It is also interesting to note that Ina's mother was illiterate in both English and Chinese and indeed never learnt to speak the English language. In contrast, Ina's Australian-born father was schooled in both Australia and China in the first two decades of the twentieth century. The language and educational differences between Ina's parents, itself an outcome of gender and class unevenness, was perhaps an underlying factor in this division of labour and space (as also found by Song 1995 in the British context).

References

Bagnall, K. 2004, '"He Would Be a Chinese Still": Negotiating Boundaries of Race, Culture and Identity in Late Nineteenth Century Australia', in S. Couchman, J. Fitzgerald,

P. Macgregor (eds.) *After the Rush: Regulation, Participation and Chinese Communists in Australia 1860–1940*, Otherland Press, Fitzroy, VIC, pp. 153–170.

Blunt, A. and Dowling, R. 2006, *Home*, Routledge, London.

Blunt, A. and Rose, G. (eds.) 1994, *Writing Women and Space: Colonial and Postcolonial Geographies*, The Guilford Press, London.

Choi, C. 1975, *Chinese Migration and Settlement in Australia*, Sydney University Press, Sydney.

Commonwealth Bureau of Census and Statistics (CBCS). 1917, 'Volume II-Part VIII. Non-European Races', *Census of the Commonwealth of Australia taken for the night between 2nd and 3rd April, 1911*, Melbourne.

Commonwealth Bureau of Census and Statistics (CBCS). 1924, 'Volume.1-Part V. Race', *Census of the Commonwealth of Australia taken for the night between 3rd and 4th April, 1921*, Melbourne.

Commonwealth Bureau of Census and Statistics (CBCS). 1937, 'Volume.1-Part XII. Race', *Census of the Commonwealth of Australia 30th June 1933*, Canberra.

Delphy, C., and Leonard, D. 1992, *Familiar Exploitation: A New Analysis of Marriage in Contemporary Western Societies*, Polity Press, Cambridge.

DeVault, M. L. 1991, *Feeding the Family: The Social Organization of Caring as Gendered Work*, The University of Chicago Press, Chicago.

Finch, J. 1996, 'Women, "The" family and families', in T. Cosslett, A. Easton, and P. Summerfield (eds.) *Women, Power, and Resistance: An Introduction to Women's Studies*, Open University Press, Philadelphia, PA, pp. 13–22.

Fitzgerald, J. 2007, *Big White Lie: Chinese Australians in White Australia*, University of New South Wales Press, Sydney.

Fitzgerald, S. 1997, *Red Tape, Gold Scissors: The Story of Sydney's Chinese*, State Library of NSW Press, Sydney.

Gandhi, L. 1998, *Postcolonial Theory a Critical Introduction*, Allen Unwin, Crows Nest, NSW.

Giese, D. 1997, *Astronauts, Lost Souls & Dragons: Voices of Today's Chinese Australians in Conversation with Diana Giese*, University of Queensland Press, St Lucia, Queensland.

Goffman, E. 1959, *The Presentation of Self in Everyday Life*, Doubleday, New York.

Greer, G. 1970, *The Female Eunuch*, MacGibbon & Kee, London.

Greer, G. 1999, *The Whole Woman*, Doubleday, London.

Huck, A. 1968, *The Chinese in Australia*, Longmans, Croydon, VIC.

Hugo, G. 2002, 'Effects of international migration on the family in Indonesia', *Asian and Pacific Migration Journal*, vol. 11, pp. 13–46.

Inglis, C. 1972, 'Chinese in Australia', *International Migration Review*, vol. 6, no. 3, pp. 266–281.

Kamp, A. 2018, 'Chinese Australian women's 'homemaking' and contributions to the family economy in White Australia', *Australian Geographer*, vol. 49, no. 1, pp. 149–165.

Kamp, A. 2021, 'Chinese Australian Daughters' Experiences of Educational Opportunity in 1930s–60s Australia', *Australian Historical Studies*, pp. 1–18 (online).

Kuo, M. F. and Fitzgerald, J. 2016, 'Chinese Students in White Australia: State, Community, and Individual Responses to the Student Visa Program, 1920–25', *Australian Historical Studies*, vol. 47, no. 2, pp. 259–277.

Lam, T., Yeoh, B. S. and Law, L. 2002, 'Sustaining families transnationally: Chinese-Malaysians in Singapore', *Asian and Pacific Migration Journal*, vol. 11, pp. 117–144.

Li, P. S. 1993, 'Chinese investment and business in Canada: Ethnic entrepreneurship reconsidered, *Pacific Affairs*, vol. 66, no. 2, pp. 219–243.

Ling, H. 1997, 'A history of Chinese female students in the United States, 1880s–1990s', *Journal of American Ethnic History*, vol. 16, pp. 81–109.

Martínez, J. T. 2021, 'Mary Chong and Gwen Fong: University-Educated Chinese Australian Women', in K. Bagnall and J. T. Martínez (eds.) *Locating Chinese Women: Historical Mobility between China and Australia*, Hong Kong University Press, Hong Kong, pp. 204–229.

May, C. 1984, *Topsawyers: The Chinese in Cairns 1870 to 1920*, James Cook University History Department, Townsville.

Mohanty, C. T. 1984, 'Under Western eyes: Feminist scholarship and colonial discourses', *Boundary 2*, vol. 12, no. 3, pp. 333–358.

Ramsey, C. 1986, 'Mostly celebration - a student who stayed', in M. Loh and C. Ramsey (eds.) *Survival and Celebration: An Insight into the Lives of Chinese Immigrant Women, European Women Married to Chinese and their Female Children in Australia from 1856–1986*, Published by the editors, Melbourne.

Rolls, E. C. 1996, *Citizens: Flowers and the Wide Sea*, University of Queensland Press, St Lucia, Qld.

Silvey, R. 2006, 'Geographies of gender and migration: Spatializing social difference 1', *International Migration Review*, vol. 40, no. 1, pp. 64–81.

Song, M. 1995, 'Between "the front" and "the back": Chinese women's work in family businesses', *Women's Studies International Forum*, vol. 18, no. 3, pp. 285–298.

Song, M. 1997, 'Children's labour in ethnic family businesses: The case of Chinese take-away businesses in Britain', *Ethnic & Racial Studies*, vol. 20, no. 4, pp. 690–716.

Spender, D. 1980, *Man Made Language*, Routledge & Kegan Paul, London & Boston.

Stacey, J. 1976, 'A feminist view of research on Chinese women', *Signs*, vol. 2, no. 2, pp. 485–497.

Summers, A. 1975, *Damned Whores and God's Police: The Colonization of Women in Australia*, Penguin Books, Ringwood, VIC.

Teng, J. E. 1996, 'The construction of the "traditional Chinese woman" in the Western Academy: A critical review', *Signs*, vol. 22, no. 1, pp. 115–151.

Watson, R. S. and Ebrey, P. B. (eds.) 1991, 'Marriage and inequality in Chinese society', *Studies on China*, 12, University of California Press, Los Angeles, CA.

Williams, M. 1999, *Chinese Settlement in NSW: A Thematic History/a Report for the NSW Heritage Office*, NSW Heritage Office, Parramatta.

Wilton, J. 1998, 'Chinese stores in Rural Australia,' in K. L. MacPherson (ed.) *Asian Department Stores*, Curzon, Richmond, Surrey, pp. 90–113.

Wilton, J. 2004, *Golden Threads: The Chinese in Regional New South Wales*, New England Regional Art Museum in association with Powerhouse Publishers, Armidale/Sydney.

Wolf, M. 1972, *Women and the Family in Rural Taiwan*, Stanford University Press, Stanford.

Zarrow, P. G. 2005, *China in War and Revolution, 1895–1949*, Routledge, London.

5 Cultural maintenance in homes and families

Throughout the White Australia Policy era, the segregation and assimilation of those considered beyond the national imaginary functioned to fulfil the national desire for a racially and culturally homogenous 'White' society. Those who were not physically excluded from the Australian nation 'were required to abandon their distinctive cultural values, lifestyles, customs, languages and beliefs and conform to the national [Anglo-Celtic] way of life' (Haebich 2008: 12). Acceptance (or tolerance) in the wider (White) community could only be achieved in this way (Wilton and Borworth 1987; and Tan 2001). Through the lens of Chinese Australian female lives, in this chapter, I will demonstrate that within this hostile and challenging context, the manner in which Chinese Australians responded to the pressures to assimilate varied. Some Chinese Australians largely abandoned 'old ways' in favour of assimilation into dominant White society, while others carefully adapted; negotiating between links to the familiar—'home' and Chinese cultural practices—and conforming to new 'Australian' ways of life in order to obtain a sense of national belonging. The latter suggests the presence of cultural pluralism long before the advent of formal multicultural policy (in the 1970s) and challenges essentialist notions of 'White Australia'. Also importantly, interviews with Chinese Australian women pointed to the relationship between gender, class, geographical location and generation on such experiences of cultural maintenance or abandonment.

In this chapter, I will also, again, draw attention to the domestic realm and illustrate that homes and family institutions were key sites in which cultural practices were maintained. I will demonstrate that females, particularly mothers, were important and central players in the maintenance and reproduction of those cultural practices[1]. In other words, females can be viewed as the primary bearers of culture or 'cultural custodians' within Chinese Australian homes and families in the White Australia context. This contributes to postcolonial feminist research, particularly emerging from the discipline of sociology and geography, on the role of females as cultural reproducers in immigrant communities and the role of 'home spaces' as sites of diasporic identity (re)construction (see Sen 1993; Yuval-Davis 1997; Dasgupta 1998; Ebaugh and Chafetz 1999; Le Espiritu 2001; Lee et al. 2002; Tolia-Kelly 2004; Long 2013; Morrice 2017; and Ratnam 2018, 2020). These findings also provide important historical, geographical and intersectional contributions to existing research regarding the frequency

DOI: 10.4324/9781003131335-5

of participation in cultural activities and perspectives on cultural maintenance among Chinese Australians in more recent times (see Dunn and Ip 2008).

Despite an emphasis on the links between cultural maintenance and women's domesticity, the findings presented in this chapter (like Chapter 4) challenge Orientalist and often structuralist-inspired feminist understandings of Chinese women's universal oppression and victimisation within the domestic sphere. I suggest that by maintaining links to 'home' and reproducing Chinese cultural practices, albeit in the confines of households and family institutions, Chinese Australian women resisted White assimilation. Therefore, in line with the works of hooks (1991), Honig (1994), Young (1997), and Legg (2003), I provide further evidence of the liberating potential of 'home' for women in oppressive societies.

Chinese Australian families and the context of isolation

Like other Chinese diaspora settlements, the nineteenth-century settlement of Chinese in Australia saw the establishment of 'Chinatowns'—hubs of commercial, cultural, social and political activities for Chinese migrants and their descendants. In Sydney, the 'Chinese Quarter' was initially located at The Rocks from the 1850s (moving to the Haymarket area in the early twentieth century where it continues to exist) and in Melbourne, in Little Bourke Street—now the heart of a much extended Chinatown area. In and around these locations, Chinese Australian businesses, places of employment and residences existed side by side, even throughout the White Australia Policy era. In these spaces, Chinese Australians were able to converse in local dialects within a social context, obtain food products and kitchen utensils imported from China or fresh produce grown by local market gardeners, and continue religious practices in temples and other places of worship. They were also able to actively participate in community organisations (Tongs) based on kinship and clan ties, assert their ties to China through political organisations such as the Guomindang (KMT), and keep up to date with current affairs in China via a variety of Chinese language newspapers (Teo 1971; and Kuo 2013)[2]. For those Chinese Australians living in suburbs or regional towns, the ability to maintain cultural practices in social settings would have been more difficult (as noted by Wilton 1998).

The childhood lives of all the Australian-born participants (as well as Patricia who migrated with her family at the age of six, and Helen who migrated as a teenager to complete her high school education), was marked by cultural isolation and minimal (if no) contact with the central Chinatown hubs. Doreen recalled that growing up on the southern outskirts of Sydney, in the 1940s and 1950s was a time in which 'there were only two or three other Chinese families in the local area'. Kaylin, who also lived in the suburbs of Sydney, recalled a sense of separation from the Chinese Australian community:

> One of the unusual things about our family, I think, was that whereas most of the Chinese in Sydney of those days lived in the city where … most of the Chinese worked around the area. We always lived in the suburbs so that we were, in a way, removed from most of the Chinese community.

Other participants who grew up in regional areas also spoke of their sense of isolation from other Chinese. Mei-Lin, who grew up on the New South Wales far north coast, explained: 'we were so isolated from other Chinese families. The closest would have been forty miles away and we had cousins sixty miles away'. A sense of isolation was most commonly recalled in regards to school life, with many participants explaining that they were the only or one of very few Chinese pupils in their primary and high schools. Such recollections included those of Australian-born Eileen who grew up in far north Queensland:

> Once again we went to school—there was a majority of say Aussie people. Gee, there was hardly a Chinese face there. So we mixed with the Aussies. You had the occasional Aborigine [sic] or the children from the Torres Strait Islands. But once again very, very simple—played with the Australian friends at playtime.

Helen, who migrated to Australia from Hong Kong and boarded with an Irish Australian family while attending an all-girls Catholic school in Sydney, had similar memories:

> I was the only Chinese girl in my year at the end. [...]. There was another one who was local born but I never had any association with her because she was a few years younger.

These recollections of being one of very few Chinese Australians in neighbourhoods and communities hint at what can be termed 'double isolation'. That is, participants were not only culturally isolated because of policies of assimilation and discrimination, but because they were physically removed or distanced from others in similar circumstances. In this way, the opportunity to maintain culture—language, traditions, celebrations—in social contexts was limited.

Despite this 'double isolation', most of the interview participants did recall that Chinese culture was practised within families in the privacy of homes. This parallels assertions made by Teo (1971) and Inglis (2011) that for most Chinese who did not live in Chinatown, social life was centred on family and friends. In the interview participants' cases, the ability and desire to maintain cultural practices were defined by parents' wishes and cultural knowledge. For some participants, 'Chinese culture' and 'Chineseness' was explicitly encouraged and enforced by parents. For the majority, however, culture was maintained in the routine realities of everyday life—eating Chinese food and speaking Chinese to parents were common experiences in the childhood home. In adulthood, many participants continued this parental guidance of Chinese cultural maintenance within their families. In this way, the family was the unit of 'cultural transmission' (Anthias 2000: 16). These aspects of cultural maintenance in homes and families are presented in the remainder of this chapter.

Language

'Australianness' has (not unproblematically) been defined, in part, by notions of an intrinsic and shared cultural identity and set of values rooted in an Anglo-Saxon heritage (Gilroy 1987; Hage 1998; Shaw 2006; Haebich 2008; and Kamp 2010). Central to 'Australian culture' is the English language, with its proficient use by individuals within the national borders an indication of rightful belonging in the Australian nation and successful integration into Australian society (Haebich 2008: 168). In the White Australia context, demands for immigrants to abandon their native languages (at least in public) in favour of English were entrenched in political and public discourse. Those that did not, often experienced racial taunts, discrimination and exclusion (Haebich 2008).

For many Chinese immigrants in nineteenth- and twentieth-century Australia, the English language was not easily or readily adopted. As I detailed in Chapter 1, the majority of Chinese immigrants came from rural areas of Guangdong Province in southern China. As such, they spoke various dialects of Cantonese that were particular to their villages and clans. The Australian-born interview participants (and New Zealand-born Daphne) were descendants of such migrants and Chinese-born participants were also from Cantonese speaking regions: Hong Kong (Sandy, Mary and Helen), Jiangxi Province, north of Guangdong (Lily) and Dongguan district in Guangdong Province (Patricia). Wilton (2004) has argued that the majority of early immigrants did not possess the skills to read or write Chinese fluently and had limited English language abilities. There were some individuals within Chinese Australian communities that did, however, learn to speak English and had various levels of English literacy. Some became fluent and acted as interpreters, while for others 'the English vocabulary acquired was limited to the words needed to seek employment and conduct business with English-speaking employers and customers' (Wilton 2004: 51).

Interestingly, 1911 census data on literacy levels indicate a relatively large number of Chinese Australian females who were proficient in the English language—a stark contrast to the largely illiterate Chinese immigrant community described by Wilton (2004). In fact, the majority of Chinese Australian females in 1911 could read and write English (see Table 5.1). There were differences between the literacy levels of 'full Chinese' and 'mixed Chinese' females: 78 per cent of 'mixed Chinese' females could read and write English compared to 40 per cent of 'full Chinese' females. 'Full Chinese' females were, however, more likely to read and write a foreign language only (10 per cent) compared to the negligible number of 'mixed Chinese' females that had that skill. The difference between the two groups may be reflective of the different levels of exposure to English/Chinese language in homes (with 'mixed Chinese' females assumedly having one English-speaking parent) and/or the fact that many 'full Chinese' females may have been immigrants with little English-language education. The census report of 1911 does not indicate how many females could read and write English *in addition to* a foreign language. This information would be helpful in understanding the bilingualism of Chinese Australian females.

Table 5.1 Literacy of Chinese Australian Females, 1911

Literacy	*'Full Chinese'*[1] (%)	*'Mixed Chinese'*[1] (%)	*Total* (%)
Read and Write English	363 (40)	1,127 (78)	1,490 (64)
Read only English	2 (0.2)	0 (0)	2 (0)
Read and Write Foreign Language Only	92 (10)	1 (0)	93 (4)
Read Foreign Language Only	7 (0.8)	0 (0)	7 (0.3)
Cannot Read	333 (37)	304 (21)	637 (27)
Not Stated	100 (11)	6 (0.4)	106 (5)
TOTAL	985 (100)	1,537 (100)	2,335 (100)

1 For the years 1901–1954, figures for 'Full Chinese' have been appropriated from the category 'Full Blood Chinese' and 'Mixed Chinese' from the category 'Half-Caste Chinese' as defined by the Commonwealth (later Australian) Bureau of Statistics.

Source: CBCS (1917).

The language skills of interview participants paralleled the patterns in the census data, with all the women interviewed—migrant and Australian-born—being fluent English speakers. As Bagnall has noted, such language proficiency suggests cultural competency among Chinese Australians and integration into 'mainstream' Australia (Bagnall 2004: 160; see also Wilton 2004). Several participants were also fluently bilingual (Lily, Sandy, Mabel, Doreen, Mary, Patricia and Helen) and others had various levels of Chinese language skills.

While 'ancestral language [...] is usually the first trait which disappears in the process of assimilation' (Teo 1971: 590), Cantonese dialects did not disappear from Chinese Australian communities. Given the limitations of the census data, one tangible piece of evidence for this language maintenance was the existence of Chinese language newspapers. With the majority of the Chinese Australian population being literate and increasingly bilingual in the inter-war years, in the early 1920s up to five Chinese language weekly newspapers were in print (Jones 2005; MOCAH 2005). Beyond these public demonstrations of Chinese language maintenance, little is known of the ways in which language was practised within the more private and intimate sphere of homes and family contexts in the White Australia era or, more specifically, women's experiences of Chinese language use.

The role of mothers in language maintenance

Participants' recalled that the maintenance of Chinese language in childhood homes was not usually a result of enforced learning/use of the Chinese language by parents, but an 'incidental' consequence of their mother's own language ability. With mothers' responsibilities often centred on the running of the household and care of children (or at the 'back' of family businesses), opportunities to learn English were limited and it was generally unnecessary as interactions with the broader Australian community were rare (as also argued by Wilton 1996). As such, despite policies of assimilation, migrant mothers were often unable to speak English. This was also partly a consequence of the intersection of migrant mothers' gendered, cultural and economic/class positions. Several did not have the benefit of formal education. As Doreen recalled: '[m]y mother spoke a pigeon English but was illiterate because she was just a little village girl in China and she didn't even have much education.' The limited language skills and restricted educational opportunities of migrant mothers contrasted with that of fathers' who had greater educational opportunities and bilingual skills. For example, Ina and Nancy's migrant mothers had limited English language skills and education. On the other hand, their Australian-born fathers were sent back to China to complete their education—a common experience among Chinese Australian families who sought a 'decent Chinese education' for their children, particularly sons (Fitzgerald 1997: 51; see also Bagnall 2004). Patricia's father, who was born in China, also benefitted from a thorough education. Patricia recalled the distinct gender divide between her parents' educational achievements:

> [Mother] said to me herself that she would have liked to have had an education and she would teach herself certain words and learn just enough to write and just to read a bit. [...] Whereas my father had a deep knowledge of the classics and all that because he was sent to be educated, you see.

Therefore, several participants had to speak Chinese in order to communicate with their migrant mothers. This inadvert learning of Chinese often resulted in broken or 'pigeon' Chinese language skills, which meant that participants were never fully conversant. As Ina recalled: 'I never learnt to speak Chinese brilliantly but I did speak because my mother couldn't speak English'. Similarly, when I asked Patricia if she spoke Chinese in her childhood home she replied with:

> Oh, yes, to mum. Because she never spoke English, you see. She never knew how to, so she always spoke to us in Chinese. But the Chinese we spoke was very general. We didn't have deep conversations, I think maybe because I was studying all the time.

By correlating her time spent studying with her inability to 'have deep conversations' in Chinese, Patricia also suggested that her formal (English-language) education was prioritised over her maintenance of Chinese language. Her relationship with her mother and their language barriers were mirrored in Nancy's interview:

> I always felt that [my mother and I] had a pretty close relationship. Even though, for all those years, she never learnt English and I never became fully

fluent in Chinese, so it's always a regret of mine that we could never really talk about the 'deep and meaningfuls' because of the language differences.

Therefore, in the very act of 'doing gender' (Ebaugh and Chafetz 1999: 588), some participants' mothers were unintentionally 'cultural custodians' of language. With their roles and responsibilities centred on the domestic sphere, learning English was not necessary for communication in public contexts. In this way, migrant mothers not only maintained Chinese language skills but passed them down to their children within the safe environment of the home. Although their maintenance of Chinese language was not a conscious political act against discriminatory policy, these examples reveal that some migrant mothers played important and central roles in families' resistance to assimilation. This highlights a complex interplay between women's patriarchal oppression (in being relegated to the home and having minimal educational opportunities) and empowerment within an oppressive context.

The important role migrant mothers played in passing down language skills was also highlighted in participants' recollections of their mother's passing away. For example, Sally linked her mother's death to a loss of language skill:

> We were very lucky there because my mother couldn't speak English, very little. She actually said in English, 'Sorry, I don't speak English.' So we were very lucky there, that we spoke Chinese. But unfortunately, as years went by, my mother died and being educated in the western schools and having a lot of Australian friends, I rarely speak Chinese unfortunately, but I do understand it. I can say a few phrases but that's about all. But the culture is still very strong.

Sally's mother was the 'custodian' of Chinese language as a direct consequence of an inability to speak English. Sally perceived this in a positive light, claiming she was 'lucky' because her mother's limited English language skills facilitated her own learning of Chinese. In this way, Sally placed great value on the maintenance of Chinese language in her family context while overlooking any possible disadvantages her mother may have encountered due to her language skills. Although she can no longer speak Chinese, Sally somewhat defended or legitimised her own sense of 'Chineseness' by asserting the strength of her Chinese culture (see also Ang 2001).

In contrast, for some second and third generation Australian-born participants, Chinese languages were absent in childhood homes and families because Australian-born mothers could not speak Chinese. Edith explained that even though her Australian-born father spoke Chinese, it was her mother's lack of Chinese language skills that impacted on her own inability to speak Chinese:

> In my father's family we were the only ones that didn't speak Chinese because my mother didn't speak Chinese because when she was young her mother was in the town running a business and her father was not at home [...]. She helped in the shop from a young age so they all spoke English.

Edith's experience again indicates that language was passed down through the maternal line. Similarly, Mei-Lin's Australian-born mother could not speak Chinese, thus impacting her own language ability:

> Mum didn't speak any, not much. She said she had a year when they were little that they went to China and when they came back she spoke nothing but Chinese, but that left them soon after. So no, we didn't speak Chinese.

While the link between Mei-Lin's mother and the family's abandonment of Chinese language is clear in this excerpt, I will discuss in the following section that Mei-Lin's inability to speak Chinese was also greatly influenced by her Australian-born father who, while having Chinese language skills, refused to speak it.

In their own adult lives, interview participants seemed more consciously active in passing down Chinese language skills to their children and resisting assimilation. Australian-born Stella, for example, has limited Chinese language skills yet, with the help of her husband, attempted to teach her children the language: 'We tried to teach them. They all know a few words, mainly food ingredients, that's all'. Lily, who migrated with her husband and two eldest children from Hong Kong in 1971, established a demerit system to encourage her children (including an additional two Australian-born) to speak Chinese: 'We tried very hard to keep [the Chinese language] but after the children went to school they tried not to speak Chinese so we had a system of fining them five cents if they spoke English'. Many participants tried a more formal approach, attempting to send children to Chinese language classes. Sandy, in fact, went so far as to run a Chinese language school: 'because I wanted them to learn Chinese'. In this way, Sandy attempted to take her role as cultural custodian beyond the confines of the home and family institution.

Despite their desire to pass on language skills to their children, these attempts were generally unsuccessful. A primary reason for the loss of language among the younger generation was strident resistance from children. As Doreen recalled:

> When I said to the kids, 'I want you to go off and learn Cantonese', they wouldn't have a bar of it. As far as they were concerned they were surrounded by mostly Australian friends.

Doreen's recollections suggest that her children's resistance was based on their belief that learning Chinese was unnecessary as their peers all spoke English. It also hints at notions of 'belonging' based on cultural similarity—language allowed Doreen's children to 'belong' in their group of 'Australian' friends. Belonging to another group, culturally via language, was not their priority. This link between rejection of Chinese language and sense of belonging to the broader Australian community was reiterated by Patricia. She explained her children's refusal to speak Chinese in the following way:

> They don't even know what you're talking about if you speak Chinese. [...] When I sent them to Chinese school, they didn't like it. They wanted to be in the crowd.

These experiences indicate that the maintenance, or rather prioritisation, of Chinese language was indeed problematic in the Australian context and came at the price of potential isolation and segregation from peers and mainstream society for children.

Some participants also indicated that English provided greater communication opportunities for their children and was a more reliable and efficient language used in the home. Lily, who began fining her children if they did not speak Chinese, soon changed her perspective and opted to encourage English as a consequence of communication barriers. She explained:

> In the end we couldn't really continue [enforcing Chinese language] because we found that if you continue that—and there are many things they can't express in Chinese—then you lose communication with your children. It's better to have communication with them rather than strictly enforce the language thing.

Lily therefore prioritised active communication with her children over Chinese language maintenance. In a similar way, Patricia sent her children to Chinese language school, yet admitted that they were largely unsuccessful due to her and her husband's own limited language skills and the ease of speaking English in the home:

> Because my husband spoke English all the time at home and myself, he knows a bit of Chinese and I knew a bit, but not deep. Not enough to talk about news bulletins and intellectual things. So we all spoke English because communication was much faster. By the time we'd rack our brains to find the right word, the whole day is gone. So we always spoke English.

Therefore, unlike her own mother who could only speak Chinese, there was no imperative for Patricia's children to learn the language.

The maintenance and teaching of Chinese language can be seen as an empowering resistance to assimilation. However, participants did not overtly characterise their experiences in this way. Rather than a political act, some participants explained that the passing down of Chinese language skills was simply rooted in the desire for children to maintain a sense of connection to the past and to communicate with relatives. Edith explained that she sent her children to Saturday Chinese school because she 'knew that they had to have a connection to their grandparents' and because of this she 'tried to keep up those traditions so that they learn that side of the culture'. Similarly, although Patricia was largely unsuccessful in teaching her children Chinese language, she believed it was an important means for her children to maintain connections to extended family:

> I felt they will never be able to communicate with their cousins in China if they don't know the language. So that's why I thought it may be a good idea to send them to Chinese school.

Patricia also drew upon notions of identity and belonging in explaining her conscious efforts to ensure her children learnt the Chinese language:

> Also when I took them to Chinese school, another important factor was that there were other kids like you. You're not the only one growing up in Strathfield like that, isolated, because I felt that when I was young. I had a great problem with identity. Why are all these beautiful people with curled eyelashes and I don't have curled eyelashes and beautiful skin and all that. I had that feeling of identity—lost identity. So in taking the kids to Chinese school, there were other kids the same, exactly the same.

For Patricia, assimilation could never be a reality as her physicality always separated her from the White Australian majority. Therefore, Saturday school provided a means in which her children could learn the Chinese language and provided a space in which they could feel a sense of 'belonging' as a result of shared physical and cultural identity.

Fathers, language and identity

In contrast to other participants' lives, for Mabel and Doreen, childhood was a time in which Chinese language was explicitly encouraged or enforced by their fathers. They recalled that this was intrinsically linked to their fathers' own sense of 'Chineseness' and hope that Chinese identity was passed down through the generations. When I asked Mabel whether she spoke Chinese at home in her childhood she responded by explaining: 'Yes, we all spoke Chinese. I think my father in particular encouraged it. He said, "If you want to eat rice, if you want to eat, then you should speak Chinese"'. Mabel's father was Australian-born and asserted his Chinese identity and sought to maintain close connections to his Chinese cultural heritage. By directly associating Chinese food culture with Chinese language, he implied that the eating of rice is a marker of 'Chineseness' and the speaking of Chinese a validation of that Chineseness. The maintenance of Chinese culture and identity within Mabel's home was therefore imperative and provides a pertinent example of the way in which language is an important marker of cultural/ethnic identity and a central tool for maintaining links to 'homeland' among migrant communities more broadly (Gudykunst and Ting-Toomey 1990; Miller and Hoogstra 1992; Hurtado and Gurin 1995; and Phinney et al. 2001).

Mabel's Australian-born father asserted his right to Chinese identity alongside Australian citizenship. Perhaps, in this way, he attempted to recreate a 'home away from home'—passing down skills he had learnt from his parents and creating a sense of the familiar through the maintenance of language and food culture in Australia. While Doreen's migrant father also maintained the Chinese language at home, it was not to recreate a sense of 'home', but as a consequence of his sojourning mentality. Doreen explained:

> So after the war there were three Australian-born children but through that period my father had all intentions of taking us back to China. And so, until

I was five I spoke nothing else but this village dialect of Cantonese so that we would still fit in when we went back to the village.

Language was thus a practical necessity for Doreen's father's plan to return to China: '…while my father was reasonably well educated in Chinese, he could understand English but he didn't want to learn English because he felt he was going home'. Assimilating and 'fitting' into broader White Australian society via language was therefore not a priority.

In Doreen's childhood, language was not only a reflection of her (and her father's) Chineseness but also a reflection of her classed identity. She recalled that her father was insistent that his 'village dialect of Cantonese' was maintained, rather than a 'proper' form of Cantonese:

> There was extended family, but of course my mother or even my father didn't like it because my great aunt spoke proper Cantonese where my family were like country bumpkins and my father likened it to putting on the talk and being snobbish. And he didn't want us to learn her kind of dialect. He wanted us to have the village bumpkin dialect.

Doreen's recollection suggests that there was not a single 'Chinese identity', but multiple forms of Chineseness based on socio-economic position. This importantly points to difference and diversity within the Chinese Australian community and within Chinese Australian families. It also indicates that Chinese Australians, like Doreen's father, consciously asserted their identity via their choice of dialect and used language as a means to differentiate themselves from others in the community and within the more intimate sphere of extended family.

The correlation between language and identity was similarly revealed in recollections of participants whose parents actively rejected the speaking of Chinese in the home. Mei-Lin explained that in addition to her mother's inability to speak Chinese, her own lack of Chinese language skills was linked to her father's overt decision to abandon the language. She explained:

> No they didn't [speak Chinese]. And I used to say to my father, 'cause his mother and father didn't speak any English, and I used to say, 'Well why didn't you speak Chinese? Why didn't you keep your Chinese?' And he'd say, we were Australian … I think he might've had a harder time. They might've had a harder time. He didn't talk about it but I'm sure they did and he maintained that no, we're here in Australia so therefore we will be Australian.

In a similar way, pressures to assimilate were also felt by Eileen and meant that Chinese language was not maintained and practised in her childhood home. When I asked Eileen if her parents thought it was important that she retained the Chinese language she recalled:

> No, not at all, because in the early days you didn't have the freedom like today. […] [Speaking Chinese] was seen as being very rude. You don't speak another language except English.

Within this context of cultural intolerance, Eileen's mother 'had to learn to speak English' during her earlier settlement in the tropics of Northern Australia. This in turn impacted Eileen's own ability to speak Chinese: 'So she [mother] thought no point of teaching the children Chinese because we've got to speak English. She didn't know the future would bring much more freedom'. While Eileen's experience of assimilation and abandonment of Chinese language paralleled that of Mei-Lin, her recollections are particularly unique as they indicate that it was her mother, not her father, who decided the family would forego the Chinese language. This experience is in further contrast to that of the other participants; her mother's role in teaching language was not inadvertent but decidedly cognisant. This points to a more formal role of Chinese Australian women to pass down cultural knowledge within the family institution.

Like participants' children who rejected speaking Chinese in order to 'fit into' Australian society, Mei-Lin, Kaylin and Eileen's recollections of their parents' abandonment of Chinese language points to the pressure to choose between two separate identities—being 'Chinese' or being 'Australian'. Their recollections of childhood language maintenance suggest that there was no possibility of an alternate subjectivity which incorporated both. Bagnall (2004), in her examination of mixed Chinese families in colonial Australia, highlighted similar notions of discrete and mutually exclusive identities and argued that the choice to maintain an 'Australian' identity came at an assumed cost: 'the loss for many of connections to China and the Chinese community, the loss of family tradition and name and culture' (p: 167). Indeed, this sense of loss was communicated by Mei-Lin, Kaylin and Eileen in regards to language. However, in the following section and Chapter 6, I will illustrate that this sense of loss was not all-encompassing and the clear distinction between the two cultural identities was not a reality for many participants. Some participants who did not speak Chinese maintained a sense of Chinese cultural identity in other ways. Some also claimed a sense of both Chinese and Australian identity and negotiated the constructed boundaries between the two.

Food culture and 'Chineseness'

Unlike language, which we have seen is often abandoned through the generations, food habits are generally one of the last cultural practices to be lost in contexts of migration and ethnic minority cultures (Tuomainen 2009; see also Charon Cardona 2004; Spiro 1955; and Teo 1971). This was certainly the case for many participants' families in White Australia. Even for those participants who did not speak Chinese and claimed an overriding 'Australian' identity, food was an important and everyday means of maintaining some link to Chinese cultural identity. For example, Mei-Lin and her family did not speak Chinese within the home and abandoned most aspects of Chinese culture: 'basically we were Australian' (Mei-Lin), yet eating Chinese-style meals was maintained: '[W]hen we were younger we'd used to say, 'Why do we have to be Chinese? Why do we have to eat all this rice? Why can't we have steak and eggs?' Therefore, although Mei-Lin explained that in most aspects she was very much 'Australian', she believed that food culture, centred on the eating of rice, was an important marker of her

Chinese identity and differentiated her from the broader White Australian community. In her view, there was a distinct division of what defined 'Chinese' food and what defined 'Australian' food—with rice certainly not deemed part of the Australian diet.

Mei-Lin's understanding of the relationship between food (particularly rice) and identity reflects a 'dominant food ideology' (a phrase used by Charles and Kerr 1988) shared by other interview participants. As Sally and Eileen remembered:

> We always had white rice, that's the basic Chinese meal in our family and a lot of Cantonese families. Not fried rice, plain rice and two dishes: a simple dish of beautifully cooked steamed fish or a dish of steak and Chinese vegetables.
>
> (Sally)

> The majority was Chinese food because with the boiled rice, you ate simply. Mum grew a vegetable garden at the back. Then she had her own chooks. If they went fishing they brought a fish home. [...] Food was just very, very simple.
>
> (Eileen)

Chinese food culture and eating habits were not only characterised by the type of produce used and the way they were prepared and cooked, but also by eating etiquette and eating habits (Wilton 2004: 75; and Shun Wah 1999). This included using particular utensils (chopsticks and bowls) and eating as a group which was important as 'it provided a time for bonding and sharing' (Wilton 2004: 74). In Doreen's recollections of her family's food culture, the continuation of such eating habits and etiquette in the White Australia context were illustrated:

> So there were all of us crowded around this table and eating very, very different kinds of food, [...] everything cooked with rice [...] and even the fact that we were eating with chopsticks where other places were with knives and forks. And [...] if dad had a good season we would come into Chinatown and have our own private room and have a ten course banquet and that happened quite frequently, probably four or five times a year but we'd always be in a Chinese restaurant and again, the chopsticks and everything else like that.

Unlike other interview participants who did not have the opportunity to frequent Chinese restaurants and cafes, Doreen and her family were able to visit Chinatown in Sydney to celebrate her father's economic successes with a 'ten course banquet'. For Doreen, going to Chinatown for these celebratory meals was again a reflection of her cultural identity.

Like Mei-Lin, Doreen also asserted that her family's food culture was a point of difference from mainstream Australian society. The above recollection was prompted when I asked Doreen to explain her childhood home life to me. She premised the recollection by stating: '[w]ell, I was always ashamed to bring anyone home because it was such a primitive place'. Doreen then directly linked this sense of shame to her food culture:

> And [my friend Nancy] could smell the Chinese cooking and she was hoping I would say, 'would you like to stay and have dinner?' And instead that never happened because I was so ashamed about the way we ate and everything else.

In Doreen's view, her family's food culture was something to hide and generated a sense of distance between her and her White Australian peers.

Despite the feelings of shame and embarrassment experienced by participants such as Doreen and Mei-Lin, Chinese food culture has been one of the primary means in which cultural divides between Chinese and non-Chinese Australians have been bridged (Wilton 2004). An example is Chinese tea which has played an important role in Australia's food heritage and cross-cultural contacts. In the nineteenth century, the importation of tea from China was one of the first tangible links between China and the colony. It became the key beverage consumed by all Australians and its centrality to cross-cultural contact was epitomised and symbolised in the Sydney tea salons of the famous Chinese merchant, Quong Tart (Wilton 2004). Chinese food culture slowly entered the Australian diet. From the mid-twentieth century, Chinese restaurants which were frequented by both Chinese and non-Chinese Australians emerged in Australian towns and suburbs. Sally's father's Modern China Café in Sydney and Eileen's parents' noodle shop in Cairns are testament to the adaptation and adoption of Chinese food culture in the twentieth-century Australian context. With Chinese and non-Chinese customers, these 'eating locales' were not only 'important means of cultural identification and enactment of the past' (Tuomainen 2009: 528), but a means in which Chinese Australians could share their food heritage. Chinese cooking styles have also entered Australian homes. Mary's recollection of visits to friends' houses in her teenage years provides insight into the various culinary manifestations of Chinese food. She recalled: '[My friends'] mothers always insist[ed] on cooking fried rice and it was awful! I had to eat all their horrible pretend Chinese food. They're being kind of course, but they had no idea how to cook'. Despite her reluctance to eat the 'Australianised' versions of her native dishes, Mary's experience illustrates the way in which Chinese food has played a key role in creating spaces and places for cultural exchange (Wilton 2004: 83) and the contributions of Chinese Australians to Australia's food heritage.

Mothers and the maintenance of food culture

As I discussed in Chapter 4, the preparation of meals in homes was generally the responsibility of wives and other female members of families. This functioned as an important means to maintain Chinese food cultures in the White Australia context. With migrant mothers most often cooking Chinese-style meals for their families, and with limited opportunities to eat Chinese food in restaurants or cafes, women and homes were pivotal to the continuation and passing down of food culture and knowledge. Eileen recalled that despite the limited produce available, her mother was extremely resourceful, preparing *ham yee* (salt fish), a traditional Chinese ingredient that could not be easily found in Australian shops or markets. Eileen's recollection of this food tradition was raised in the interview when I asked her if her non-Chinese childhood friends ever saw any differences between Eileen and themselves. In response, Eileen explained:

> Maybe a few little things they would kind of laugh at, but kind of brushed off. Like the salt fish—which is yummy—and mum would make salt fish. The house was built up on large post type of thing because the air circulation and all that, the Queenslander. Mum would have these salt fish hanging down from the beams. Kids would come and, what's that, gee it smells, what's that?

While Eileen spoke proudly of her mother's continuation of her family's food heritage via ingenious methods to provide her family with authentic Chinese ingredients, she perceived that this food culture was a cause for criticism and mockery from non-Chinese Australians. In this way, her food culture placed her outside the Australian 'norm' and marked her difference: '[w]e thought of it as nothing. But to them, that was something different to them'. Mei-Lin similarly recalled her mother's cooking of *ham yee* as a point of difference from the broader Australian community:

> My mother used to like *ham yee*, Chinese smelly salty fish. And if ever she got it and cooked she'd say, 'Oh, I hope no one comes to the front door! Oh I hope they don't come!' And we'd say, 'What's the matter?' And she'd say, 'Oh they'd think we're terrible with all these terrible smells!' But no, we didn't. There weren't that many Chinese restaurants then and the Chinese food you got in Chinese restaurants was terrible! So you know we wouldn't… if we ever went I'd be likely to have steak and eggs rather than Chinese food!

Like Eileen, Mei-Lin's mother believed that cooking *ham yee* reflected on her badly: 'Oh they'd think we're terrible'. Yet despite this fear of a negative response, Mei-Lin's mother maintained her food culture and made important contributions to her family's sense of cultural identity. This recollection also reiterates that home was the only place in which Mei-Lin and her family could enjoy their native cuisine as there were very limited options to eat Chinese food in restaurants. Mei-Lin's home was a crucial site in which her family could maintain their food culture and her mother, in her role of feeding the family, was inadvertently positioned as the 'keeper of culture'. Through a postcolonial feminist lens, this role cannot simply be viewed as part of the process of patriarchal oppression, but more complexly as a role which empowered Eileen and Mei-Lin's mothers to maintain cultural practices and identities despite broader contexts of assimilation and intolerance.

Mothers also played a central role in maintaining food culture by passing down cooking skills and knowledge to the younger generation. Susan's mother taught her Chinese cooking skills: '[Mother] taught us how to do *dim sums* and things like that, and we watched her as she did her own cooking'. Susan, however, explained that she did not maintain the skills her mother taught her as the pressures of her adult life saw her losing interest in her Chinese culture:

> I don't really cook much Chinese food. I'm terrible, [...] well after I lost my husband I think I lost interest in… I mean I had to provide for the children of course, but, I don't know, we just didn't seem to have much Chinese food except if we went to my mother's place or if she came over and brought food with her.

By referring to her abandonment of Chinese cooking as 'terrible', Susan implies that maintaining Chinese food culture was something that she *should* be doing. Due to limitations of the interview material, it is not clear as to why Susan perceived her food culture in this way. Her recollection does, however, further highlight the central and important role of her mother as the keeper of culture within her extended family context—Susan's mother was the only link Susan and her children had to their Chinese food traditions.

Unlike Susan, other participants continue to maintain cooking skills they learnt as children in the White Australia context in their current adult lives. When I asked Mei-Lin if she cooks Chinese food, she explained:

> Yes I do! I must say when I've been overseas on a holiday the first thing I do when I come home is go and buy roast duck and cook some rice and do a proper Chinese meal.

Mei-Lin resented the fact that she was Chinese and had to eat rice all the time in her childhood: 'We'd go past the houses and smell the steak and onions and we'd be having rice *again*. [...] I remember saying, "*Why* do I have to be Chinese?!"', however, as an adult, Chinese food provides Mei-Lin with a sense of comfort and familiarity. Nancy similarly continues to cook Chinese food and as a mother herself has passed on her food culture: 'it's one legacy that I like to think I'm leaving to our children'. Without the ability to speak Chinese, Nancy perceived her ability to cook Chinese food as an important and crucial link to her Chinese heritage. By describing her food culture as a 'legacy', Nancy highlighted the important role of food in multi-generational cultural knowledge, awareness and maintenance. It also suggests that she has adopted the role of cultural custodian in her adult family context and makes important contributions to her family's continued maintenance of Chinese cultural traditions. Eileen similarly reiterated this notion of food as a cultural legacy and her own role as cultural custodian. While she asserted that she 'cook[s] everything', it is her Chinese cooking skills that have particular pertinence in her family:

> But I say to my kids, 'look you enjoy the Chinese food, you always come back and say, oh mum, can you cook this dish or this Chinese dish'. I said, 'look you kids better marry a Chinese person because when I'm gone you won't get it'. [They say] 'Oh don't worry, we can go to the Chinese restaurant or you can teach them.

It is important to note that while Eileen's children have the opportunity to eat Chinese food in contexts beyond the home (unlike the previous generation) they still ask Eileen to cook them traditional meals. Eileen also recognises that she is an integral component of this aspect of cultural maintenance: '...when I'm gone you won't get it'. As such, in feeding her family, she continues her maternal role as keeper of culture and, like Mei-Lin and Nancy, makes important contributions to the maintenance of Chinese food culture within homes in contemporary Australia.

Although the majority of meals in participants' childhood homes were in the Chinese style, an adoption of more Western cooking styles was not uncommon. Eileen, who remembered her mother preparing *ham yee* also recalled her mother and aunt adopting more Western cooking styles:

> Dad's young sister married an English solider and so she was cooking more Aussie meals because she worked as a housemaid, so I suppose she learnt to cook more the Aussie type foods. On the farm [before she was a housemaid], aunty cooked more the Chinese food, very simple Chinese food. After a while Mum began to cook a bit of everything.

Once again, this experience emphasises the contributions Chinese Australian women made to the running of households and maintenance of culture. It also illustrates the importance of home as a place in which culture could be maintained. In Eileen's aunt's experience, home '[o]n the farm' was a place in which she cooked Chinese food. In contrast, when she began her employment as a housemaid—assumedly in an Anglo-Australian household—she was exposed to more western-style cuisine and consequently 'learnt how to cook more the Aussie type of foods'. While Eileen's recollection highlights the difference between the types of food cooked in Chinese Australian households and Anglo-Australian households—hinting at a perception that there is a discrete division between 'Chinese' and 'Australian' food cultures—it also indicates that Chinese Australian women were able and willing to adopt alternative cooking styles and 'cook a bit of everything'.

The influence of the broader social context on food culture and the willingness of Chinese Australians to adopt alternative cuisines are also highlighted by Stella's recollections of growing up in the multicultural context of Thursday Island. Brought up by her migrant grandmother, Stella recalls that her family ate a variety of foods rather than just Chinese cuisine:

> On Thursday Island [...] one lot of people took from the others and so there's almost a distinctive Thursday Island type of food. A lot of its Malaysian, the Chinese introduced rice, the Japanese...my friend's the Yamishuta's, the father was a soya sauce maker and we cooked a lot of Malayan type food. Plus we had traditional English food like roast beef and corned beef.

For Sandy, it was simply too difficult to maintain Chinese food culture due to the limited produce available. Therefore, unlike the majority of other participants, she and her sister prepared mostly Western-style meals:

In those days it was not easy getting anything Chinese. So all the vegetables are broccoli, beans and the stuff like that and the meat will be cut the Australian way, so we mainly had steaks and chops.

These excerpts indicate a willingness of Chinese Australian women to learn and embrace western food culture and in the majority of cases, do so while still maintaining their own Chinese food traditions. Such cultural fusions/dynamisms can be viewed as an important contribution to the development of a unique 'Chinese Australian' culture, or indeed, 'Australian' culture which, through food, embraces a multiplicity of traditions.

Beyond the everyday: rituals and seasonal festivities

The maintenance of formal Chinese rituals and practices, experiences were similarly diverse. Some participants, like Kaylin (whose family abandoned the Chinese language), recalled that her family did not celebrate formal Chinese festivities because they were isolated from the Sydney Chinese community: '[…] we didn't go through all the Chinese festivals and things, you know, they still have in Sydney. I don't know it might've been because we lived so far out of the city'. Hong Kong-born Mary, who spent her adolescence in boarding school in Maitland in northern New South Wales, similarly explained that she was unable to continue any Chinese traditions after her arrival in Australia: 'No, not at all. How can it be? There's just nothing'. In her interview, Mei-Lin also explained that practising Chinese traditions and festivities (beyond the eating of Chinese food) was not possible:

There weren't that many [traditions] that you could maintain.

(Mei-Lin)

What about Chinese New Year?

(Alanna)

Not really because there we were up in the country. It was really only when you were in Sydney that you could get to see it and we also… now perhaps in country towns… I don't think they do in country towns. They don't have the dragons and things like that. But, no basically we were very much Australian.

(Mei-Lin)

As Dunn and Ip (2008) noted in regards to the links between community cultural maintenance and the presence of cultural infrastructure among contemporary

Chinese Australians, Kaylin, Mary and Mei-Lin's recollections suggest an intrinsic link between cultural maintenance (in the form of participation in community events) and geographical location. For these participants, their location in areas where few other Chinese Australians resided and distance from major cities like Sydney where Chinese Australian communities were established and cultural infrastructure had been developed meant that the ability to maintain culture in a community setting was limited.

In the above recollection, Mei-Lin also highlighted the intrinsic link between cultural practice and identity. While she associated her food culture with her Chinese identity, in this instance she asserted that her and her family's inability to celebrate more formal Chinese cultural events such as New Year meant that they 'were very much Australian'. Thus, for Mei-Lin, a loss of cultural practice was equated with a loss of cultural identity. Despite this assertion, in a recollection of her childhood in Texas, Queensland, Mei-Lin remembered dressing in traditional Chinese costume for the celebration of Empire Day:

> I remember we dressed up one [year]. It must have been the end of the war when we all got into our little Chinese costumes, our little cheongsams. I have a photo of us being in a parade. That's all I remember. But there was no sort of friction because of the other Chinese families there.

This experience of dressing in *cheongsams* (see Figure 5.1) is significant in a number of ways. Firstly, although Mei-Lin explained that it was difficult to participate in community celebrations of Chinese festivities such as New Year's Eve due to the cultural isolation, this experience indicates that more personal practices of Chinese culture were possible. Thus, Mei-Lin's assertion of loss of Chinese culture was perhaps overstated. Secondly, by participating in the parade dressed in traditional Chinese costumes, Mei-Lin's experience suggests a negotiation of Chinese and Australian identities/cultural performance. Empire Day, on the 24th of May, was an annual celebration established in honour of Queen Victoria and celebrated throughout the British Empire. By participating in the parade dressed in *cheongsams*, Mei-Lin and her family celebrated the British Empire while maintaining aspects of their traditional Chinese culture. Mei-Lin asserted that this merging of Chinese and Australian cultural identity was not negatively received by the broader community. This points to an assumption that such a declaration of 'Chineseness' would not normally be tolerated.

Figure 5.1 Dressed in traditional Chinese costume for the Empire Day Parade, Texas, Queensland, c. 1945. In this photograph, three-year-old Australian-born Mei-Lin Yum (right) stands with Nancy Lee and Joyce Lee at the Empire Day Parade.

Source: Mei-Lin Yum (private collection).

Other participants experienced more strident efforts to maintain seasonal Chinese festivities and rituals in their homes and families—most notably, rituals associated with *Bai San* (ancestor worship) such as those conducted during *Qingming* (*Ch'ing-ming*) as well as the celebration of Chinese (Lunar) New Year. These practices and celebrations came in a variety of forms as they were adapted to the specific Australian environment and in relation to participants' families' particular social and economic situations.

Celebrating Chinese New Year

Chinese New Year's Day (or the Spring Festival) is celebrated on the first day of the first lunar month which, according to the Gregorian calendar, falls in late January or early February each year. In the Chinese calendar, it is the most important date of the year and is marked by various celebratory practices (Huang 1991). Central to the celebrations are family banquets on New Year's Eve, the giving of offerings to ancestors at the family altar, further family reunions on New Year's Day, and the visiting of neighbours, relatives and friends. Gifts are exchanged, usually in the form of sweets or red packets (*lai see*) filled with money, and entertainers are often called upon to perform the lion dance and sing good wishes (Huang 1991).

Interviews with Chinese Australian women provided insights into the ways in which Chinese Australian families celebrated Lunar New Year in the White Australia context. While the festival's traditional emphasis on family was reflected in participants' experiences, unlike the colourful and elaborate festivities described by Lin (1960), Mabel and Patricia explained that their family celebrations were a modest event:

> Maybe at Chinese New Year maybe my mother might've cooked a little more but in the early days we weren't all that well off so we wouldn't have had roast duck for example.
>
> (Mabel)

> We had Chinese New Year, but it wasn't a big thing. It was maybe just a few special foods and people give you little red packets, you know what they are. There's money in it. So some people would give you a few dollars or even one dollar would be—well in those days it was pounds—in pounds, shillings and pence. Very simple and not grand like some people would have it.
>
> (Patricia)

Mabel and Patricia indicate that their New Year celebrations were limited by their economic circumstances. Nonetheless, their families maintained whatever practices they could despite official policies of assimilation. Sally, on the other hand, experienced more elaborate Chinese New Year celebrations:

> We always celebrated Chinese New Year. That was a huge occasion [...]. You had to do a thorough house clean and have a special Chinese dinner with fried rice and special dumplings, and we had to wear nice new clothes. [...] [W]hen the friends visited they gave the daughters [...] little Chinese red packets which was really nice.

The difference between Sally's experiences of Chinese New Year, and those of Mabel and Patricia, was probably due to the better economic situation and general social status of her family—particularly Sally's father, Hung Jarm Pang. Not only did Hung Jarm Pang run the Modern China Café (a prominent business in Sydney's Chinatown in the 1940s and 1950s), but he was also Chairman of the Sydney Sze Yap Society (also known as Sydney Sze Yap Wooi Kwoon)[3]. Thus while other participants recalled a sense of cultural isolation during their childhood, Sally's childhood years were marked by interactions with the broader Chinese Australian community in Sydney, connections which, as her memory of friends visiting suggests, allowed more social New Year celebrations. Commonalities were, however, seen across the economic and social divide. Most participants recalled receiving *lai see* and Sally and Mabel both remembered eating special foods during the season (an example of maintaining food culture beyond the everyday experience). However, regardless of economic standing or social situation, it is clear that Chinese Australian families resisted assimilation and were able to maintain the basic elements of the Chinese New Year season within the White Australia context. This can be viewed as an important contribution to the cultural fabric of modern multicultural Australia in which Chinese/Lunar New Year celebrations are a hallmark occasion—particularly in Sydney where celebrations have drawn millions of spectators and participants each year.

Women's memories of childhood celebrations of New Year, like memories of the maintenance of language and food culture, pointed to the central and important role of mothers in the seasonal festivities. Margaret recalled the efforts of her mother in keeping the tradition alive: 'Mum used to try and do the customs like try and celebrate Chinese New Year and stuff like that and have the food and things'. Doreen similarly recalled the central role of her mother in New Year celebrations:

> There'd be no way she'd let any of us wash our hair on Chinese New Year's day, how we'd just got castigated if we broke anything, or if we burnt anything, how the house had to be all cleaned by New Year's Eve, how all the debts had to be paid. So she stuck to those old Confucian ideas. So there was a basis, it wasn't just superstition.

Both Margaret and Doreen made no mention of their fathers in their reflections of New Year celebrations. In this way, they constructed their mothers as the 'keeper of culture' in their childhood homes and families. Margaret's father died when she was approximately 15 years of age and this could have been why she remembered her mother single-handedly maintaining the New Year traditions. Doreen, however, grew up in a 'nuclear' family context. For Doreen, her mother's role as cultural custodian was intrinsically linked to her mother's gendered roles and responsibilities in the domestic sphere. Her overseeing of her children's hygiene and the cleanliness and maintenance of the house during the New Year period—all reinforced Doreen's mother's gendered role within the patriarchal family system. In her own adult life, Doreen attempted to continue this role, wishing to pass down New Year traditions to her sons. However, with her family unable to see the 'logic' of the New Year rituals, her mother's 'very strict superstitions went by the wayside' (Doreen).

Nancy, however, also highlighted the role of both her mother and father in New Year festivities. She explained: 'I vividly recall Dad would make sure all his bills were paid before Chinese New Year and Mum would have special food for us'. While both of Nancy's parents participated in the customs of the season, the gendered divisions between Nancy's mother's and father's roles during this time were clear. Nancy's mother also contributed to the New Year's celebrations through the sharing and handing down of homeland memories:

> But it was a time that she [mother] really talked about her childhood and she'd say how it was…she loved Chinese New Year because she'd get a new dress and how there were crackers and people would go from house to house and she'd receive a lot of lai-see […] and they flew kites. So it was nice to hear her talk about that and she'd talk about it even years later. They were very happy memories for her.

In this way, Nancy's mother's role as 'cultural custodian' went beyond the maintenance of the family's traditional food culture. She passed down and educated her children about the traditions she experienced as a child in China and played an emotive role in the Chinese New Year festivities. This sharing of memories perhaps allowed a continued sense of connection to China and espoused a sense of cultural identity.

Ancestor worship

Ancestor worship has been a central component of Chinese cultural practice as far back as the Shang dynasty (1600–1050 B.C.) when '[b]oth real and symbolic gifts of elaborate burial goods, human sacrifices, and offerings of food and clothing provided the deceased with the comforts of life' (Rouse 2005: 20). Confucianism, with its emphasis on ritual, reinforced the importance of rites associated with ancestor worship (Rouse 2005: 20). These rites included the observance of a number of holidays to honour the dead, including *Qingming*—a time marked by the cleaning of ancestral graves, burning of incense and the offering of food and money for ancestors to use in the afterlife (Rouse 2005). Families would also continue honouring the deceased in homes and ancestral halls through the ancestral tablet—a wooden plaque displaying the name, rank and title of the deceased. Underlying the practices of ancestor worship is the belief that the dead can influence the lives of the living. Thus, '[p]roper ritual following the death of a relative or friend was essential not only to the soul of the departed but also to the happiness, harmony, and well-being of those left behind' (Rouse 2005: 19). In line with Confucian ideas of social harmony, family hierarchies and filial piety, it is also believed that the dead are superior to the living. As such, the deceased must be given the utmost respect and reverence (Huang 1991).

The centrality of ancestor worship to Chinese cultural life was reflected in the practices maintained by Chinese immigrants and their Australian-born descendants in nineteenth- and twentieth-century Australia. The practice of exhuming the deceased from Australian burial sites for their return and burial in ancestral homelands was common and important among Chinese Australians and other overseas Chinese communities (Fitzgerald 1997; and Williams 1999). Despite the extent of the practice[4], returning bones was not mentioned by interview participants. This was perhaps because the practice was in decline after 1930—the period in which most participants spent their childhood.

Interview participants did, however, recall other practices associated with ancestor reverence, particularly participation in *Qingming*. When I asked Sally if she participated in the *Qingming* ritual she replied: 'Yes, we did definitely. I remember that vividly. We went to Rookwood cemetery to visit my sister's grave and we'd take the incense and the chicken. It was an offering to the loved ones.' Marina similarly remembered:

When my mother died, we always used to have to go to the grave on special *Qingming* time, which is about twice a year … and we had to go and take food and rice and chicken and pork and wine and crackers and … paper and all that. Then we had *Bai San* […], worship the dead. We used to do that a lot. That's a very…a sort of tradition that not many people in Australia do. The old people do.

Chinese ancestor worship has traditionally reflected the patrilineal family system in that it is the forebears on the father's side, particularly male forebears, which must be venerated (Walstedt 1978: 389; see also Das Gupta et al. 2003). Sally's and Marina's experiences of honouring the graves of female members of the family during *Qingming* indicate variation on this tradition. Marina's explanation that worshipping the dead is no longer a common tradition in Australia also highlights generational differences between cultural practices and the important role she and her family had in continuing the ancestor rituals.

After her father's death, Ina also participated in ancestor veneration rituals. Unlike Sally and Marina who mainly visited the cemetery once or twice a year at *Qingming*, Ina, along with her sister, visited her father's grave every week. When I asked her if she practised ancestor rituals throughout her childhood, Ina recalled:

> Yes I did. Because my father died—you'd only practise these traditions if you had someone dead in the family [...]. We used to go every Sunday with my older sister driving. But she wouldn't go, my mother, because [...] she grieved in the Chinese style which is very open, emotional, almost out of control grieving.

The 'Chinese style' grieving referred to by Ina was a common practice at funerals in nineteenth- and twentieth-century China (Rouse 2005). The loud wailing of family (and often hired professionals) demonstrated their extreme grief and mourning (Rouse 2005). Ina and her sister's weekly visits to their father's grave and their mother's continued grieving suggests that for Ina's family, the maintenance of traditional death rituals was a priority. In fact, this prioritisation is further reflected in Ina's continued practice of the more formalised annual *Qingming* ritual. She explained her maintenance of the ritual in the following way:

> When we go to Rookwood once a year to clean the graves, and it's always in the month of April, [...] belief has it that this is the month that all the souls of the long dead come down to visit the people and the gates of heaven are open [...]. And we put food on the grave and we pray.

During this festival, Ina plays an important role in passing down knowledge of the ritual to her sons. She explained:

> Now I have to tell my sons, 'Right now, pick up the whiskey – two hands, two hands – pick up the whiskey and do it like that', and they're just obeying instructions. And when I think of their children, they will be again, one generation removed. And it just becomes watered down.

This explanation not only highlights Ina's role as cultural custodian within her family context, but also suggests a belief that Chinese cultural knowledge, within the Australian context, will naturally be lost through the generations as children/grandchildren become increasingly distanced from herself—the 'keeper of culture'. This sense of loss (and the important role of women in cultural maintenance) was reiterated by Margaret: 'well [when] my father passed away we used to do *Bai San*—to the special ancestry thing that you had to do for that. I try to keep it up, but the kids, they don't do it'. Once again, Margaret suggests a loss of cultural knowledge and practice through the generations.

Adopting 'Australian' traditions

Just as interview participants' mothers embraced and adapted to European cooking techniques and ingredients, interview participants also recalled an adaptation to 'Australian' traditions and festivities such as Christmas and Easter. Like food culture, language and traditional Chinese rituals, these more 'Western' traditions were also celebrated in the domestic sphere and led by participants' mothers. Patricia recalled:

> When Christmas came along, well she [mother] tried to...oh, we're almost celebrating Christmas or we'll have a special lunch, mainly at home. Again, maybe something a little bit special, but nothing grand. [...] So it wasn't a big thing and my...I suppose because of the economic situation, we didn't go into it very big.

Patricia asserted that she has no religious affiliations and that her parents were, in fact, Buddhist/Taoist. Due to limitations in the interview data, it is not known why Patricia's family practised the Christian celebration despite their adherence to an alternative religion. Beyond the religious incongruence, Patricia's experience is significant as it provides another example of the importance of the home in cultural practice. Her recollection also reinforces the notion that women, particularly mothers, are responsible for maintaining cultural traditions in the home. Interestingly, while immigrant women's roles as 'cultural custodians' typically refer to the maintenance of cultural practices from 'home'—immigrants' place of origin—in Patricia's case, the cultural practices maintained by her mother were not only those from her homeland, China, but also those from her 'host' country, Australia.

As a migrant mother herself, Lily recalled her take-up of Christmas traditions after her arrival in Australia in addition to the maintenance of Chinese festivities:

> We celebrate our Chinese New Year and all the festivals, the lantern festival we went to the park with them and other children, yeah we tried. But also we celebrated Christmas even though we're not Christians. My explanation is because you know, everyone else is celebrating it, and if they're left out they [my children] feel isolated.

Lily's recollection is pertinent as it illustrates an adaptation to the Australian context and negotiation of cultural identity that saw both 'Chinese' and 'Australian' cultural practices coexisting and being practised in tandem. For Lily, the uptake of Christmas was not related to religious belief but was, instead, a means in which she could obtain a sense of belonging to the broader Australian community.

Conclusion

Utilising a postcolonial feminist approach, this chapter once again illustrates the diversity of experience among Chinese Australian women. Home, particularly in participants' childhood lives, provided a central and important space in which Chinese cultural practices were maintained and cultural knowledge was passed down through the generations. However, experiences of cultural maintenance in homes and family contexts were varied. In some instances, the maintenance of Chinese cultural practices was not a priority in the lives of participants. Rather, Chinese culture and identity were de-emphasised in order to conform to broader Australian society. For other participants, 'traditional' Chinese cultural practices were strictly enforced or maintained in the routine realities of everyday life. Whether or not interview participants maintained Chinese cultural practices was largely dependent on their parents' desire and ability to maintain aspects of 'traditional' Chinese life.

Language, food culture and festivities/celebrations were the primary ways parents maintained links to 'homeland' and instilled a sense of 'Chineseness' in their children. While some fathers were instrumental in the reproduction of Chinese culture in Chinese Australian homes and families, it was more commonly Chinese Australian mothers who were centrally involved in the maintenance of these cultural practices and the passing down of cultural knowledge to their children. With their responsibilities for domestic duties such as cooking, cleaning, bearing and raising children, participants' mothers were often inadvertently charged with maintaining cultural practices within the home and passing down cultural knowledge to children. By uncovering these roles and responsibilities, I have highlighted the important contributions of Chinese Australian women to the cultural fabric of their families, but also on a much broader scale, to the development of 'multicultural' Australia. In this way, I have begun to revise national discourses which have excluded Chinese Australian women and challenged essentialist notions of a 'White' Australia by indicating the presence of cultural pluralism long before the advent of formal multicultural policy.

While mothers' role in the maintenance of Chinese culture in the White Australia context is clearly linked to their position within the patriarchal Confucian family hierarchy, a postcolonial feminist perspective allows us to view this role as not one of oppression and victimisation, but of having some aspects of empowerment. Throughout the White Australia period, non-White cultural maintenance was discouraged in favour of a policy of exclusion and assimilation. By maintaining cultural practices in homes, women resisted assimilation.

Therefore, rather than viewing women's roles in maintaining culture as part of their victimised role or position within the patriarchal family, it can perhaps be viewed as part of their empowerment within a White assimilationist context. Through this perspective, we are able to further question the validity of assumptions that all Chinese women are, and have been, oppressed by Confucian ideology within family systems and by family relations across historical and geographical boundaries. Within the unique White Australia context, these Orientalist assumptions of female oppression and victimisation need to be reconsidered and redefined to accommodate the diverse experiences, actions and contributions of Chinese Australian women.

Notes

1 While the home and family unit were the primary loci for the maintenance of Chinese cultural practices, I do acknowledge that culture was maintained by participants beyond the home context in young adult friendship groups and via participation in Chinese community organisations, for example, student associations at university and the Chinese Women's Association. Participants such as Sandy, Helen, Ina and Mabel suggested that these forms of cultural maintenance were important means of (re)connecting with their Chinese culture, facilitating a sense of belonging and promoting Chinese culture to wider Australians. A discussion of these forms of 'non-familial' cultural maintenance cannot be adequately achieved within the confines of this book and I therefore suggest it is an important topic for future research.

2 Inglis (2011) has taken on a more pessimistic view. She asserted that with pressures to assimilate, second generation Chinese Australians, educated in Australia, were not literate in Chinese and 'had only attenuated contacts with Chinese culture especially after the Communist party came to power in 1949'. Additionally, while Chinese community organisations continued to exist in the post-war period (except the Chinese Chamber of Commerce which closed by 1965), 'many of the pre-war organizations were reliant on the support of a few members of the older generation'. Many of the Chinese language newspapers had also ceased publication by the 1930s. Inglis therefore argued that even in the social context of Chinatown, in the post-World War II era, 'small numbers of the Chinese population [...] made it difficult to maintain the formal organizational life and range of cultural activities' (Inglis 2011: 53).

3 The Sze Yap Society was an organisation made up of members originating from the Sze Yap area in southern China. Like other district/clan-based societies, it was established to provide support to its members, to conduct charitable work, establish temples and maintain cultural activities (Williams 1999).

4 Between 1875 and the late 1930s, 'a peak of 75% of burials in the 'Old Chinese Section' of Rookwood Cemetery were, "returned to China", with an average of 55% to 65%' (Williams 1999: 17).

References

Ang, I. 2001, *On Not Speaking Chinese: Living Between Asia and the West*, Routledge, London.

Anthias, F. 2000, 'Metaphors of home: Gendering new migrations to Southern Europe', *Gender and Migration in Southern Europe: Women on the Move*, Berg, Oxford, pp. 15–47.

Bagnall, K. 2004, 'He would be a Chinese still: Negotiating boundaries of race, culture and identity in late nineteenth century Australia', in S. Couchman, J. Fitzgerald and P. Macgregor (eds.) *After the Rush: Regulation, Participation and Chinese Communities in Australia 1860–1940*, Otherland Press, Melbourne, pp. 153–170.

Charles, N. and Kerr, M. 1988, *Women, Food and Families*, Manchester University Press, Manchester.

Charon Cardona, E. T. 2004, 'Re-encountering Cuban tastes in Australia', *The Australian Journal of Anthropology*, vol. 15, no. 1, pp. 40–53.

Commonwealth Bureau of Census and Statistics (CBCS) 1917, 'Volume i. Statistician's report', *Census of the Commonwealth of Australia*, Melbourne.

Das Gupta, M., Zhenghua, J., Bohua, Li., Zhenming, X., Chung, W. and Hwa-Ok, B. 2003, 'Why is son preference so persistent in East and South Asia? A cross-country study of China, India and the Republic of Korea', *The Journal of Development Studies*, vol. 40, no. 2, pp. 153–187.

Dasgupta, S. D. 1998, 'Gender roles and cultural continuity in the asian indian immigrant community in the U.S', *Sex Roles*, vol. 38, no. 11–12, pp. 953–974.

Dunn, K.M. and Ip, D., 2008, 'Putting transnationalism in context: Comparing Hong Kong Chinese-Australians in Sydney and Brisbane', *Australian Geographer*, vol. 39, no. 1, pp. 81–98.

Ebaugh, H. R. and Chafetz, J.S. 1999, 'Agents for cultural reproduction and structural change: The ironic role of women in immigrant religious institutions', *Social Forces*, vol. 78, no. 2, pp. 585–612.

Fitzgerald, S. 1997, *Red Tape, Gold Scissors: The Story of Sydney's Chinese*, State Library of NSW Press, Sydney.

Gilroy, P. 1987, *'There Ain't No Black on the Union Jack': The Cultural Politics of Race and Nation*, Routledge, London.

Gudykunst, W. and Ting-Toomey, S. 1990, 'Ethnic identity, language and communication breakdowns' in H. Giles and P. Robinson (eds.) *Handbook of Language and Social Psychology*, Wiley, New York, pp. 309–327.

Haebich, A. 2008, *Spinning the Dream: Assimilation in Australia 1950–1970*, Fremantle Press, Fremantle, W.A.

Hage, G. 1998, *White Nation: Fantasies of White Supremacy in a Multicultural Society*, Pluto Press, Sydney.

Honig, B. 1994, 'Difference, dilemmas, and the politics of the home', *Social Research*, vol. 61, pp. 563–597.

hooks, b. 1991, *Yearning: Race, Gender and Cultural Politics*, Turnaround, London.

Huang, S. 1991, 'Chinese traditional festivals', *The Journal of Popular Culture*, vol. 25, no. 3, pp. 163–180.

Hurtado, Q. and Gurin, P. 1995, 'Ethnic identity and bilingualism attitudes', in A. Padill (ed.) *Hispanic Psychology*, Sage, Thousand Oaks, CA.

Inglis, C. 2011, 'Chinatown Sydney: A window on the Chinese community', *Journal of Chinese Overseas*, vol. 7, no. 1, pp. 45–68.

Jones, P. 2005, *Chinese-Australian Journeys: Records on Travel, Migration and Settlement, 1860–1975*, National Archives of Australia, Canberra.

Kamp, A. 2010, 'Formative geographies of belonging in White Australia: Constructing the national self and other in parliamentary debate, 1901', *Geographical Research*, vol. 48, pp. 411–426.

Kuo, M. F. 2013. *Making Chinese Australia: urban elites, newspapers and the formation of Chinese Australian identity, 1892–1912*. Monash University Publishing, Melbourne, Victoria.

Le Espiritu, Y. 2001, '"We don't sleep around like white girls do": Family, culture, and gender in filipina American lives', *Signs*, vol. 26, no. 2, pp. 415–440.

Lee, M., Chan, A., Bradby, H. and Green, G. 2002, 'Chinese migrant women and families in Britain', *Women's Studies International Forum*, vol. 25, no. 6, pp. 607–618.

Legg, S. 2003, 'Gendered politics and nationalised homes: Women and the anti-colonial struggle in Delhi, 1930–47', *Gender, Place & Culture*, vol. 10, no. 1, pp. 7–27.

Lin, Y. 1960, *The Importance of Understanding*, The world publishing company, New York.

Long, J. 2013, 'Diasporic dwelling: The poetics of domestic space', *Gender, Place & Culture*, vol. 20, no. 3, 329–345.

Miller, P. and Hoogstra, L. 1992, 'Language as a tool in the socialization and apprehension of cultural meanings', in T. Schwartz, G. White and C. Lutz (eds.) *New Directions in Psychological Anthropology*, Cambridge University Press, New York, pp. 83–101.

Morrice, L. 2017, 'Cultural values, moral sentiments and the fashioning of gendered migrant identities', *Journal of Ethnic and Migration Studies*, vol. 43, no. 3, pp. 400–417.

Museum of Chinese Australian History (MOCAH) 2005, 'Chinese-language Australian newspapers', *Chinese-Australian Historical Images in Australia (online)*, Melbourne.

Phinney, J. S., Romero, I, Nava, M. and Huang, D. 2001, 'The role of language, parents, and peers in ethnic identity among adolescents in immigrant families', *Journal of Youth and Adolescence*, vol. 30, no. 2, pp. 135–153.

Ratnam, C. 2018, 'Creating home: Intersections of memory and identity', *Geography Compass*, vol. 12, no. 4, pp. 1–11.

Ratnam, C. 2020, '(Re)creating home: The lived and gendered experiences of tamil women in Sydney, Australia' in N. Kandasamy, N. Perera and C. Ratnam (eds.) *A Sense of Viidu: The (Re)creation of Home by the Sri Lankan Tamil Diaspora in Australia*, Springer, Singapore, pp. 119–139.

Rouse, W. L. 2005, '"What we didn't understand": A history of Chinese death ritual in China and California', in S. F. Chung and P. Wegars (eds.) *Chinese American Death Rituals: Respecting the Ancestors*, AltaMira Press, Lanham, MD, pp. 19–46.

Sen, S. 1993, 'Motherhood and mothercraft: Gender and nationalism in Bengal', *Gender & History*, vol. 5, no. 2, pp. 231–243.

Shaw, W. S. 2006, 'Decolonizing geographies of whiteness', *Antipode*, vol. 38, pp. 851–869.

Shun Wah, A. 1999, *Banquet: Ten Courses to Harmony*, Transworld, Neutral Bay, NSW, Australia.

Spiro, M. E. 1955, 'The acculturation of American ethnic groups', *American Anthropologist*, vol. 57, no. 6, pp. 1240–1252.

Tan, C. A. 2001, 'Chinese families down under: The role of the family in the construction of identity, 1920–1960', *International Conference: Migrating Identities: Ethnic Minorities in Chinese Diaspora*, Centre for the Study of Chinese Southern Diaspora, ANU, Canberra.

Teo, S. E. 1971, 'A preliminary study of the Chinese community in Sydney: A basis for the study of social change', *Australian Geographer*, vol. 11, pp. 579–592.

Tolia-Kelly, D. 2004, 'Locating processes of identification: Studying the precipitates of re-memory through artefacts in the British Asian home', *Transactions of the Institute of British Geographers*, vol. 29, no. 3, pp. 314–329.

Tuomainen, H. M. 2009, 'Ethnic identity, (post)colonialism and foodways', *Food, Culture & Society*, vol. 12, no, 4, pp. 525–554.

Walstedt, J. J. 1978, 'Reform of women's roles and family structures in the recent history of China', *Journal of Marriage and Family*, vol. 40, no. 2, pp. 379–392.

Williams, M. 1999, *Chinese Settlement in NSW: A Thematic History/A Report For the NSW Heritage Office*, NSW Heritage Office, Parramatta.

Wilton, J. 1996, *Chinese Voices, Australian Lives*, unpublished PhD thesis, University of New England, Armidale.

Wilton, J. 1998, 'Chinese stores in rural Australia', in K. L. MacPherson (ed.) *Asian Department Stores*, Curzon, Richmond, Surrey, pp. 90–113.

Wilton, J. 2004, *Golden Threads: The Chinese in Regional New South Wales*, New England Regional Art Museum, Armidale in association with Powerhouse Publishers, Sydney.

Wilton, J. and Borworth, R. 1987, *Old Worlds and New Australia: The Post War Migrant Experience*, Penguin.

Young, I. M. 1997, *Intersecting Voices: Dilemmas of Gender, Political Philosophy, and Policy*, Princeton University Press, Princeton, NJ.

Yuval-Davis, N. 1997, *Gender and Nation*, Sage Publications, London.

6 Interactions and identity

In the previous chapters, I not only indicated that Chinese Australian women were present in White Australia as wives, mothers and daughters, but were also present as active participants in public contexts such as schools and universities, family businesses and other forms of paid employment. In this chapter, I extend this discussion of Chinese Australian women's public presence by exploring individual experiences in spaces and places beyond the home. More particularly, I focus on the interactions Chinese Australian women had in public contexts (with non-Chinese and Chinese Australians) and the ways in which these interactions influenced their identities and senses of belonging. Informed by my postcolonial feminist approach, I draw upon interviews with Chinese Australian women to examine the ways in which they constructed and asserted their own identities and were simultaneously ascribed identities by others within these interactive contexts. I also examine the ways Chinese Australian women negotiated the tensions between self-ascribed and externally imposed identities and argue that various definitions and ascriptions of 'Chineseness' and 'Australianness' fostered feelings of difference or similarity. Whether self-ascribed or externally imposed, these identities were used as mechanisms for exclusion or inclusion in various understandings of race, culture and nation. By examining Chinese Australian women's accounts of their own experiences and their own perceptions of difference, identity and belonging, this chapter contributes to the recent body of literature (reviewed in Chapter 1) that aims to 'give voice' to Chinese Australians themselves. I also acknowledge the agency of Chinese Australian women in the construction of identity, their diversity of experience and the intersection of multiple axes of difference.

The discussion presented in this chapter is framed by a broader understanding that in the late colonial era and decades of the White Australia Policy period, Australia was imagined as a 'White' community—a nation defined by its racial purity, its morality, its equality and its superiority from the 'Asian hordes'. 'White Australia' was thus an 'imagined community' (to borrow from Anderson 1983), that was constructed on ideas of political autonomy, unity and identity. Central to 'White Australia' were assertions about who could and could not belong in the national space. Those of a non-White 'race' were identified as incompatible with Australia's White national identity and excluded from national belonging (Gilroy 1987; Hage 1998; Shaw 2006; and Kamp 2010).

DOI: 10.4324/9781003131335-6

Within this conception of the Australian nation, Chinese, among other non-Europeans, were (and continue to be) deemed beyond the national imaginary and viewed, via their racialised position, as 'outsiders' (Elder 2003; and Dunn 2005; Noble 2009; and Kamp 2010). As I detailed in my overview of the White Australia Policy and Chinese immigration in Chapter 1, Chinese Australians' 'outsider' status meant that they were primary targets of racism and exclusion in the colonial period and throughout the twentieth century. In fact, the presence of Chinese Australians and the 'threat' they posed to the development of the White nation—in their supposed racial inferiority, cultural incompatibility, 'contaminating' capacities and economic competition—was a primary precipitator of the *Immigration Restriction Act 1901* and the broader White Australia Policy (Kamp 2010). While many Chinese were restricted from entering the Australian nation via such legislative measures, those migrant Chinese already present in Australia and their descendants felt the brunt of the discriminatory legislation and practice in institutional and less formal settings (Palfreeman 1967; Fitzgerald 1997; Elder 2005; and Fitzgerald 2007).

As I also argued in Chapter 1, the experiences of Chinese Australians as 'outsiders' within the White Australia context has increasingly been a focus of research; however, there has been an emphasis on the male actors in Australia's Chinese past. Therefore, while the previous two chapters focused on Chinese Australian women's gendered, classed, cultural, national and 'racial' identities as developed and constructed within homes and family contexts, in this chapter, I contribute to the correction of existing 'gender blind' research by assessing how interactions between interview participants and non-Chinese Australians in the public domain contributed to participants' sense of belonging and/or exclusion. I illustrate that participants' interactions with members of the White Australian community were often marked by a questioning of the legitimacy of their 'Australianness' despite rights of birth, citizenship and generational longevity. Analysis of Chinese Australian women's recollections highlights that the primary ways in which female Chinese Australians were marked as 'Other' and beyond the national imaginary included name-calling and taunting in childhood, assumptions of foreignness, and overt discrimination and institutionalised prejudice. However, rather than take the view that Chinese Australian women lacked power and agency to assert their own identities in this context, I argue that interview participants constructed, defined and affirmed their own identities—negotiating their own national and cultural identities (along with gendered, classed and generational subjectivities) within a context of racialisation.

School-yard bullying and feelings of difference

In her examination of the childhood experiences of multi-generational Chinese Australians, Tan (2003) found school-yard taunting and bullying to be the primary means by which difference was ascribed to her respondents. Close analysis of the 19 interviews conducted for this project similarly indicated that participants' recollections of growing up in the White Australia period were often marked by memories of schoolyard taunting and name-calling which singled them out from the majority and marked them as 'different' from a White Australian norm. Sally, for example, was teased about her appearance: 'I actually had comments about my little flat nose. Boys wouldn't dance with me because I had a flat nose, but that was harmless of course'. Doreen was also taunted at school: 'at school you were obviously called names and all that, being a minority ethnic group'. Mei-Lin and Mabel, who were also Australian-born, were similarly teased at school with both participants being targets of the taunt 'Ching Chong Chinaman'. Mei-Lin explained:

> We were the only Chinese family then in that town [Ballina] […]. I didn't feel too much animosity or have too many bad things happen. But there are always children who call out 'Ching Chong Chinaman' to us which I hated! I did hate that. But it was generally from children who didn't know us. And when we had interschool visits, then you were treated rather like as if you'd come straight off from mars or you felt you did.

Although Mei-Lin recalled not feeling 'too much animosity' in her school years and claimed in her interview to have felt an overwhelming sense of Australian identity— 'We were Australian'—her account of feeling like someone from another world on interschool visits and being taunted by children who did not know her is etched in her memory as a divisive experience. The same type of enmity towards the taunt was also recalled by Mabel:

> I was always aware that I was Chinese, that I was different. And I do remember people saying, 'Ching Chong Chinaman', and… it's not the words themselves um, but if they said it sort of in a mean sort of way … naturally I could sense it and even though I was only little I used to hit them.

Like Mei-Lin's experiences, the type of 'Chineseness' ascribed to Mabel at school marked her out for ridicule and was imbued with negative connotations. As Tan (2003) has suggested, this type of school-yard ridicule 'invoked a keen sense of vulnerability among many Chinese Australians as children' and 'left an indelible mark on their sense of self and belonging' (p: 195). For Mabel, racial taunting constructed her as an outsider from a very early age.

Unlike Mei-Lin, who struggled with competing aspects of her identity—feeling 'Australian' but being ascribed 'Chinese'—in her childhood, Mabel did not claim an affinity to an Australian identity. Rather, her 'Chineseness' and difference from the White Australian norm was deeply inscribed in her sense of self. This was reiterated when she later explained: 'I knew I was Chinese and that was it. I didn't sort of have an identity problem. I didn't want to be White or anything like that'. In this recollection, Mabel's sense of identity follows Enlightenment notions of singular and stable subjectivities (as described by Cerulo 1997). Mabel's perspective assumes that 'Australian'/White and 'Chinese' identities were dichotomous and mutually exclusive and thus feeling anything but one or the other (as opposed to a combination of both identities) was a 'problem'. Later on in the interview, however, Mabel communicated a more complex understanding of her identity that challenged the binary construction of 'Australianness' and 'Chineseness'. She asserted that although she was very much 'Chinese', when she travelled abroad she felt Australian: 'You don't think of yourself as Australian unless you go overseas'. In this way, Mabel indicated that her experience of cultural/national identity was not singular or static but multiple and shifting. In this instance, she challenged Enlightenment notions of an essential, stable and unified (male) subjectivity.

With her strong sense of 'Chineseness' only felt when she was physically present in the Australian national space, Mabel's cultural/national identity was intrinsically linked to place. Mabel's identity was constructed in different ways in different (national) places—an experience that follows Pratt and Hanson's (1994) understanding of the fluid relationship between place, social process and identity. This sense of shifting cultural/national identities is a common characteristic of the transnationalism experienced by Chinese Australians in more recent times (see Dunn and Ip 2008) and, viewed from a postcolonial perspective, suggests a complex experience of place-based identity construction and negotiation.

According to Hong Kong-born Mary and Australian-born Susan, incidents of racist bullying were not common, with both participants asserting that it happened only 'once'. Mary, for example, explained:

> And when I was a student I remember once at the bus stop talking to some other students in Cantonese and someone told us to go back where we belong, because you know… 'If you want to talk, go back to where you belong', that sort of thing.

Despite the apparent rarity of the event, Mary's experience is particularly pertinent as it once again reflects the assumption that 'Chineseness' is incompatible with Australian national belonging. Perspectives on the socially constructed nature of nations assert that in addition to religion, customs or traditions and race, language is essential to national cultural identity (Gilroy 1987; Jackson and Penrose 1993; Penrose 1993; and Cornell and Hartmann 1998). As I discussed in Chapter 5, Mei-Lin's and Kaylin's fathers' abandonment of the Chinese language (in order to be 'Australian') reflected such perspectives. In Mary's account, her non-Chinese peer constructed her as 'Other' by utilising this rationale. Mary's use of the Cantonese language was viewed as 'un-Australian'—it was not part of the Australian national identity and it therefore did not belong within Australia's national borders. Her language marked her as outside the national identity.

While Mary was verbally ostracised, Australian-born Susan encountered the physical threat of stone-throwing and common taunt of 'Ching Chong Chinaman' which marked her as a racialised outsider:

> I've never really ever…except the day I started school. Someone when I was about five…I was walking home and someone threw a stone at me and said, 'Ching Chong Chinaman'. That's the only time in my whole life that I've ever felt that someone's mentioned my nationality.

In explaining this one-off incident, Susan interestingly reinforced the illegitimacy of her own belonging in Australia. Susan is an Australian citizen. She is second-generation Australian-born (her father being Australian-born and her mother an immigrant) and her family's settlement in Australia dates back to the turn of the twentieth century. Susan therefore has the birthright to claim Australian nationality. However, by describing her Chineseness as her 'nationality' Susan reiterated assumptions that her apparent 'Chinese' identity is incompatible with the Australian national identity. My analyses of Chinese Australian women's presence, experience and contributions throughout this book have challenged such understandings of Chinese Australian women's 'outsider' status and essentialist understandings of Australian identity and belonging. I have indicated ways in which Chinese Australian women like Susan indeed 'belonged' in Australia (or have a claim to 'belonging'), not only because of their presence and contributions but also through their various ways of being both 'Australian' and 'Chinese.

These experiences of school-yard taunting reiterate Tan's (2003) assertions of the vulnerability of Chinese Australian children; however, Edith's recollections of her time at school provide insight into an alternative position of power. Edith was similarly 'called names':

> [...] at school and that I'd sometimes be called names and things like that but, you know, my sister and I had this thing that we...if people picked on us we'd just pretend we could speak Chinese. So we would pretend that we would be speaking and we seemed to get over it...always get over it that way. But I yeah, I mean I know that certain things happened and it's because you're Chinese but I can't pin point anything that was... that really sticks in my mind.

Her inability to 'pin point' any specific incidents seems to downplay the significance of the school-yard bullying. However, paradoxically, it suggests that the experiences of racial taunting were not sporadic or one-off events but common 'everyday' experiences—so common, in fact, that Edith and her sister developed a strategy to deal with the racist incidents. Edith and her sister's strategy to 'pretend' to speak Chinese greatly differed from those that were used by other Chinese Australian children in the period—such as 'learning to stand up for oneself with one's fists' (like Mabel) or achieving academic success to prove equality or even superiority to White Australia peers (Tan 2003: 105). These strategies functioned as 'passports' to gaining acceptance by White Australian peers (Tan 2003). In contrast, Edith and her sister countered the taunts by actively asserting and reinforcing their Chinese identities through the 'performance' of Chineseness. In this way, they claimed 'ownership' of the externally ascribed identity rather than negating any differences.

Even though Edith asserted her power to define her cultural identity—reinforcing her 'Chineseness' rather than denying or rejecting it—this may have been due to her own perception that her 'Chineseness' was inescapable:

> I used to sometimes wonder what it would actually be like to live where you weren't different. I didn't feel it in a negative way, but just wondering what it

would be like because no matter how Australian we are [...] you will always look [Chinese] and the expectation is that you will speak [Chinese].

For Edith (like Sally and her 'little flat nose'), her Chineseness was physically inscribed on her body and contributed to her overwhelming sense of difference. This follows Tan's (2003) argument that Chinese Australian children were targets of racism as they carried 'ineradicable markers of 'Chineseness' [on] the body' (p: 102), resulting in their hypervisibility within White Australian society. In addition to physical perceptions, in this recollection, Edith made reference to the perceived links between Chinese identity and language, that is, those who are 'Chinese' are expected to speak Chinese. In making such connections, Edith indicated that her sense of difference was linked to externally imposed constructions of Chineseness.

In addition to her strong sense of 'Chineseness', be it self-defined or externally imposed, Edith also communicated a sense of 'being' Australian: 'Well I felt Chinese, but I suppose I felt Australian as well but knew I was different to other Australians. I was very aware that the perception is that you're Chinese'. Edith's sense of self was therefore complex and deviated from dichotomous understandings of 'Chinese' and 'Australian' identities. For example, school-yard taunting (and, in fact, the White Australia Policy) was based on the assumption that Chinese' and 'Australian' identities were discrete and mutually exclusive. Edith, however, felt simultaneously Chinese and Australian as a consequence of negotiating between her self-defined identity and those inscribed on her by others. Her claims to 'Chineseness', as well as 'Australianness', therefore challenged the very ideological premise of the White Australia Policy.

Participants' recollections of racial taunting and bullying at school illustrate complex power plays in processes of identity construction and ascription. On the one hand, through a postcolonial lens, we are able to view the name-calling and taunting experienced by participants in their childhood—and indeed by many cultural/ethnic groups in contemporary society (see Blair et al. 2017)—as an experience of lack of power and privilege. Just as colonisers have had the power to name and define the colonised in racialised terms, in participants' childhood lives, it was those with an inherent claim to belonging—their White Australian peers—who often designated 'Chineseness' in racialised terms and imbued it with negative connotations of difference and inferiority. On the other hand, however, interview participants also actively asserted their own 'racial' identities and in some cases challenged mutually exclusive understandings of 'Chinese' and 'Australian' identities. Furthermore, active responses to racial taunting by participants such as Edith indicate agency and an ability to shift the balance of power in the racialisation process.

Assumptions of foreignness

Besides school-yard bullying and taunting, interview participants also recalled a sense of difference and 'Otherness' via general feelings of foreignness—of belonging 'not here' but elsewhere. Kaylin, for example, was second generation

Australian-born—her maternal grandfather came to Australia in 1868 to seek his fortune on the Victorian goldfields. Despite her and her family's long history in Australia and her rights of birth, Kaylin explained that she always felt Chinese, rather than Australian: 'We always knew that we were Chinese and we were different'. For Kaylin, Chineseness and difference were synonymous. This was emphasised in her everyday experiences within the White Australian community: '[I]f you're the only Chinese in the whole suburb the eyes of everyone would be watching you if you went shopping and you did something wrong, or weren't dressed properly'. In this way, Kaylin sensed a critical gaze from the White Australian majority and felt as though she had to legitimise her presence in Australia and ensure her actions did not reflect negatively on the Chinese Australian community: '[...] and as I say, we were always conscious of the fact we must do the right thing because we would bring disrepute on all Chinese'. Not only did Kaylin feel that she was 'different' but that she was also a representative of the broader Chinese Australian community—a community that she in fact had very little contact with (as I highlighted in Chapter 5). She felt the weight of the community on her shoulders and believed she had responsibility for maintaining its reputation.

Other interview participants also recalled a sense of difference and 'Otherness' on account of their physical attributes. Daphne, who was New Zealand born and migrated to Australia in 1964, explained the way her foreignness—her Chinese identity—was externally ascribed based on her physical appearance: 'formally they would just say: "You're Chinese", right? Even though you might be a citizen and just because you look Chinese you were sort of classified as Chinese'. Margaret, who was in fact Australian-born, similarly recalled constantly being judged and treated as an 'outsider' on account of her physical looks. She explained:

> I suppose in the back of my mind I always knew I was Chinese, there's no escaping that because you looked Chinese. But you didn't sound Chinese. [... I guess if there were other people around who didn't know you, made you feel like an alien and Chinese...something different.

Like Sally and Edith's experiences of difference at school (discussed in the previous sub-section), these recollections depict experiences of racialisation where physical attributes (skin colour, hair type, eye shape, etc.) were 'racial' markers (see Tan 2006: 67). In Daphne and Margaret's experiences, their physical attributes meant that they 'looked Chinese', and as such, were 'classified' and defined as outside the national 'type'. Therefore, Daphne and Margaret experienced the power relations associated with racial classification—being defined as 'Chinese' by White Australians and subsequently inferior or outside Australian national belonging. Their experience of racialisation—where physical features 'serves as a "cue" denoting perpetual "foreignness"' (Tan 2006: 67)—are not uncommon. In the Australian context, Yang (1994), Ang (2001) and Tan (2006) have explored the tensions between 'Chinese' bodily features and notions of Australian national identity, while in the United States Yamamoto (1999) and Yu (2001) have indicated that Asian Americans encounter similar racialisation (Tan 2006).

Assumptions of foreignness were also brought to the fore when participants were asked of their origins, usually through the question, 'where do you come from?' For example, Margaret, who grew up in Sydney's western suburbs in the 1950s, recalled:

> I do remember a lot of time you get old people asking you, 'Where do you come from?' It used to irritate me so much. I thought, 'Here we go again'. You know? Because I have to go through the whole lot and say, 'I'm Australian-born Chinese'. That's what I'd say every time, 'I'm an Australian-born Chinese'.

Just as Tan (2006) noted in her exploration of the question 'Where are you from?', the questioning of Margaret's origins that arose in her interactions with older White Australians did not seem to be ill-intended. This was realised by Helen who also received such questioning of her origins: 'When people look at us, I think they more look at us out of curiosity and they will ask, "Where do you come from?" More curiosity'. Despite such innocent intentions, the question 'represents an "assumption of foreignness" by the inquirer, who assumes that since informants look "Chinese" and/or "Asian", they must be foreign to Australia' (Tan 2006: 68). In Hage's (1998) terms, the questioner can be viewed as the nationalist who does not construct the racialised nation via an 'inferiorisation or essentialisation of the other, but in the construction of the other as an object of spatial exclusion' (Hage 1998: 68). In Margaret's experience, the questioner can indeed be viewed as the nationalist who constructed and defined her foreignness; however, the interaction and relations of power were more complex. Margaret's responses to the questioning of her origins were well-rehearsed. She asserted 'I'm Australian-born Chinese'—an identity that affirmed 'belonging' in Australia via rights of birth while simultaneously declaring a Chinese cultural or ethnic heritage. Therefore, just like Edith's assertion that she felt both Chinese and Australian (discussed in the previous section), Margaret's self-affirmed identity challenged dominant notions that Australian national identity is singularly defined as White. In doing so, Margaret contributed new understandings of what it meant to be Australian in the White Australia context and asserted her own power to define her subjectivity.

Margaret similarly asserted her agency when White Australians assumed she could not speak English:

> I think we moved over to another office and this guy never knew me...what did he say about me? He said something which was rather rude. Sort of implying that I couldn't speak English or something like that, you know? And I probably said something back to him. Or [...] I think I was speaking to someone really quick because [...] I can speak English really crisp and clear, and you know. Let them know that I can speak English, thank you very much.

Despite her rights of birth, Margaret's claims to belonging and 'Australianness' were denied by her colleague—she was mistaken for a new migrant or visitor to Australia and thus assumed to be 'out of place'. In this way, her identity was 'spatially incarcerated' (Jackman 1994: 107, quoted in Tan 2006: 70) in China, a common experience for many people of Chinese descent living in the West (Ang 2001). However, like asserting that she is 'Australian-born Chinese', Margaret's reaction to speak English 'really crisp and clear' illustrated her agency to combat racialised assumptions and claim her right to 'belong' through language.

Like Margaret, Susan and Stella also explained that the legitimacy of their 'Australianness' was called into question via inquiries of their origins. Susan, who was Australian-born, explained: 'I've always felt far more Australian than Chinese, very much so. I get quite surprised when people say, "And where are you from?"' Stella, who is also Australian-born and whose grandparents were living in Australia as far back as the nineteenth century, explicitly highlighted this tension between identities, she explained: '…and I always feel that somewhere there is an inkling that I'm Chinese but it's only when it's brought to my attention'. As these memories indicate, Susan and Stella claimed a closer affinity to an 'Australian' identity throughout the White Australia Policy era. Being questioned about their origins was therefore not so much a tiresome annoyance, but a confronting reminder of their 'Otherness' and position outside the Australian national imaginary. Reiterating Tan's (2006) findings, regardless of Susan's and Sella's rights of birth, generational longevity and own sense of national/cultural identities, their claims to Australianness were constantly questioned by White Australians.

In her interview, Patricia also explained that her physical appearance marked her as different throughout the White Australia Policy era and assumptions often saw her mistaken for Japanese:

> Even in the restaurant when I was working with my brother, the customers were fine, but we were near a RSL club. They were all returned soldiers who had fought in the Japanese war, so naturally some of them would have bad feelings against Asians. When they were totally drunk and came in, they didn't know whether that was Chinese or Japanese and they would call us Japs and all that sort of thing.

Margaret's identity was similarly called into question because of her 'Asian looks':

> One of the first things they asked me was, 'Are you Chinese or Japanese?' And most of the guys that asked me this were guys who used to serve in the war, and that's why they wanted to know whether I was Chinese or Japanese. So I said 'Chinese'. Well that was ok then. I was Chinese. But if I was Japanese, God knows what would've happened.

It has been noted in existing research that assumptions of Chinese Australian's foreignness within the Australian national context often occurs alongside a collective racialisation that renders people of Asian descent (be they of Chinese, Japanese, Korean descent, etc.) as 'one and the same' (Tan 2006: 70, see also Ang 2001). This collective racialisation, as was experienced by Patricia and

Margaret, often results in a situation of mistaken identity in which individuals of one 'ethnic minority' group are confused for another or 'lumped together'. As Tan (2006), Ang (2001) and Fitzgerald (1997) have noted, this can be highly problematic—not only because it further constructs individuals as racialised outsiders, but because it 'can have harsh effects on the lives of Chinese Australians when they become scapegoats of public hostility towards other Asian groups' (Tan 2006: 73) or vice versa. According to Tan (2006: 70) and Fitzgerald (1997: 141), it was not uncommon for Chinese to be mistaken for Japanese in the aftermath of WWII and subsequently become targets of racist hostilities. In more recent times, the racialisation of COVID-19 as a 'Chinese virus' by political, media and public discourse (in Australia and globally) and the subsequent experiences of broader anti-Asian racism, has illustrated the ongoing and pervasive nature of such historic and homogenising race-based nationalism (and scapegoating) in the West (Kamp et al. 2021; and Jeung et al. 2021).

Everyday' discriminatory practices and prejudice

In Chapter 3, I discussed Chinese Australian women's 'real life' experiences of discriminatory practices associated with Australian immigration policy. While these accounts told of the formal immigration and naturalisation restrictions placed on Chinese and, more particularly, Chinese wives, it was much more common for interview participants to have experienced more 'everyday' forms of discrimination and prejudice. Like name-calling, racial taunts and assertions of their foreignness, interactions with White Australians in 'everyday' places and spaces often had real and lasting impacts on participants' sense of belonging and inclusion within their communities. For example, many participants explained that the only times in which their 'difference' from the White majority was brought to the fore was during interactions with shopkeepers and retail staff. Helen, Kaylin, Margaret and Patricia all recalled being refused service (or being served last) because of their 'outsider' status. Helen in particular recalled:

> But I did meet a few incidences in the community. Not such institutional racism, [...] I went to buy some doughnuts, would you believe. That one girl refused to serve me. She just would not serve me. It was my turn. You are in a queue, two or three people, that's all. She just would not serve me. I was standing there. So I had to walk out. That kind of [unclear] racism. But obviously because I was Chinese that she refused.

These experiences of 'everyday' racism in shops and retail outlets were therefore not uncommon and in contemporary Australia, such discrimination continues to be prevalent[1].

Beyond everyday encounters with prejudice, several interview participants also recalled discrimination within educational institutions which determined their future employment opportunities. As I discussed in Chapter 6, many of the participants recalled their gendered position as females within the Chinese family as limiting their educational achievement and employment success. These gendered limitations were not confined to the home and family but were also endemic in broader Australian social norms and views of respectability. Participants recalled that the occupations that were available and open to them, as women, were limited. For example, Margaret recalled:

> There wasn't much to do because when you were in high school we were talking about it with friends…oh, there's nothing much for girls to do. You either became a secretary or school teacher or a nurse or even a gangster's mole. That's what we came up with. And I thought, 'They're my choices', you know? So I think in the end my sister talked me into doing a secretarial course at Metropolitan Business College.

Ina, similarly recalled the gendered nature of employment:

> In those days I think girls didn't aspire to being lawyers or doctors. Most of my friends drifted really, drifted into teaching or physiotherapy or something. [...] If you couldn't get into teaching you drifted into physiotherapy and accountancy…very kind of basic degrees.

For some participants, the limitations on their education and employment were not only felt in regards to their gendered identity, but also their Chinese identity—what one participant, Lily, described as a 'double edged sword'. Patricia, for example, believed that her placement in a home economics high school, rather than a language-focused high school, may have been influenced by assumptions of her 'Chineseness'. She questioned: 'Now, is that [not being sent to a language focused school] because they thought I couldn't handle it as a new migrant not knowing English?' Patricia's perception that she was not entered into a languages school due to her 'Chineseness' parallels other Chinese Australians' experiences (see Tan 2003: 104). Tan (2003) argued that the assumption that Chinese Australian children were incapable of learning languages reflected the popular stereotype that Chinese Australians were 'inherently inferior' (p: 104) to White Australians, with the ability to learn foreign languages (English let alone French or other European languages) being beyond their intellectual capacity.

Both Ina and Mei-Lin recalled more overtly prejudiced experiences in their search for a career. Ina, who claimed in her interview that she 'melted in with the majority of Australians' during her childhood years, explained:

> I remember distinctly after graduation in Year 5 of high school [...] we all had to go to the headmistress, Mrs Whitehead, to tell her what we want to do. And I remember distinctly saying that I wanted to be a teacher. And I can recall that she said, 'But you are Chinese?' Now, I just took that, 'Well, ah gee that's a bit of a … I wonder if I can still do it'. But she didn't say you can't do it, but she said that in the tone as if, but you're blind, or you've only got one leg…how can you do that?

From Ina's point of view, her 'Chineseness' was perceived by her teacher as a disability—a physical characteristic which made her incapable of successfully entering a teaching career and thus, like Patricia, her 'Chineseness' was constructed as a marker of her inherent inferiority. Mei-Lin experienced similar discrimination when pursuing her teaching career, she recalled:

> It turned out that it was because I was Chinese that they hadn't granted me a college scholarship because apparently [...] young Chinese women who had gone into teaching had been experiencing trouble in the schools with some of the students [...]. But when I said, 'Oh, but I want to do Early Childhood, I want to be an infant school teacher', he sort of went, 'Oh...' and then so he changed his mind. And then I went on to teach for forty years and I didn't have any trouble with anybody.

Unlike Ina's experience, Mei-Lin's initial rejection from Teachers College was not based on an assumption that as a 'Chinese' individual she lacked the ability to be a teacher but was, instead, based on a paternalistic rationale—it was assumed that because of Mei-Lin's Chinese background she needed to be protected from a potentially racially hostile environment. Mei-Lin, however, proved the department wrong, achieving a successful career in education.

Once in the workforce, some participants felt that their Chineseness and their gendered identity intersected in ways that fostered feelings of differentiation. This was particularly the case for Margaret and Stella who were employed in White, male-dominated workplaces. Margaret, for example, worked in the Australian Tax Office from the late 1960s and when I asked her if she felt particularly Chinese in her interactions in that social context she explained:

> I still felt Chinese. But mixing in the tax office now is different; there's quite a few Chinese working there now. Before when it was just me, I did feel Chinese-y. I did feel I was different, you know? [...] Not only was I Chinese but I was female as well, you know?

Margaret's feelings of difference were thus rooted in her position as both a female and a Chinese Australian in the White, male-dominated world of taxation. In this way, her subjectivities, like Lily's, proved to be a 'double edged sword'. In a similar way, Australian-born Stella was employed in the newsroom of the Australian Broadcasting Corporation from the age of 17 in 1948 until her retirement in 1996, and despite feeling very much Australian, recalled the difference she felt in the early years of her employment:

> Well you always do [socialise with colleagues] in a newsroom and they're always drinking and carrying on. That's where I struck quite a bit, I think it was very strange really because with people like journalists who you'd think would be more worldly, they just didn't know how to accept the fact that they had a little Chinese working there. They honestly didn't know. And I think the most common question I got asked was how do you cook fried rice. I could hardly cook at all! It really used to throw me.

Just as the question 'Where are you from?' is generally not ill-intended (Tan 2006), Stella's experience of being asked 'How do you cook fried rice?' did not have malicious intent. Nonetheless, the questioning not only marked Stella as foreign, but assumed that she fulfilled her 'traditional' gender roles in the home.

Intersections of 'race' and class

The feelings of 'difference' experienced by many of the interview participants were not solely rooted in their racialised Chinese identities. As I discussed above, their gendered positions as females also impacted their experiences as 'outsiders' within the broader White Australian community. Additionally, classed/socio-economic positions intersected with notions of Chineseness to form yet another layer of 'difference' within White Australian communities. This intersection of Chineseness and socio-economic difference always occurred on account of participants feeling that they were of a lower class than their White Australian peers during their childhood. Doreen, for example, recalled:

> We were always dressed in rags, you know. Whenever we came home from school we had to take our school clothes off and put on hand-me-down and rags and what-not. I can understand that other Australian families probably thought they wouldn't want their children playing with us.

Doreen went on to explain that her and her family's low socio-economic position and outsider status was exacerbated by their Chineseness:

> But then going to a Catholic School, I can remember there were low socio-economic families there. [...] So there were other families that were probably of a similar socio-economic group but because we were Chinese we were different. Poor, Irish Catholic type families were probably more accepted no matter how low their socio-economic group was. [...] They were not seen as much a threat as what the yellow peril might've been [...]. We were different in how we looked, different in [...] what we ate.

Patricia had a similar experience, feeling shame and embarrassment of her family's low socio-economic situation:

> Another young lady that I was friends with—the Greek friend—did ask me to her place once. I thought I'll do the right thing and ask her back. I asked her to my place first and then I went to her place and of course her place, being at Maroubra, was a very nice house. So after that, I said I'm not asking her to my place again.[...] I was too class conscious. I don't know why.

During her interview, Patricia mentioned feeling isolated during her childhood growing up in Sydney. When I asked her what was the cause of this isolation, she responded by explaining:

Well everything, I think, Alanna, because you get only English everywhere. You get magazines and everything in English and you get beautiful film stars and everything and all the people around you are Caucasian. […] You see and you read about all the things they have […] and you feel I don't have those. You feel isolated although your parents keep telling you, 'Work hard and then you can buy it yourself'. So there was a bit of feeling of isolation like not having the same. So people I think generally feel that and I'm no different. I felt that and going to their homes, I see the beautiful things they have and felt a bit, I don't have it.

With little representation of 'Chineseness' in mainstream society, the 'White' norm dominated Patricia's everyday life—in terms of English language and Caucasian faces. Her sense of difference was further exacerbated by the socio-economic disparity between herself and her White Australian peers (pictured in Figure 6.1). According to Patricia, such differences in language, appearance and class led to feelings of isolation during her childhood years in the White Australia Policy era.

Interestingly, according to Patricia, her socio-economic position also marked her difference within the Chinese community in Sydney. When I asked her if she participated in Chinese community events such as social gatherings and debutante balls, Patricia responded by explaining: 'No, because we weren't financially recognised, I suppose. I know they did have it. […] It just depended on where you were in the economic times'. In a similar way, Mabel recalled her low socio-economic position as a point of difference within her Chinese Australian communities. In her recollections of her attempt to 'be a Christian' in her young adulthood, Mabel remembered the class consciousness she felt when attending a Chinese Church:

Oh, actually at one stage I was trying to be a Christian. It must've been when I started university, or maybe high school, no it must've been first year university or something. For a brief time I probably thought maybe I you know, I should be a Christian. Though, I went along to Church a few times but I was conscious that the people there were rich um, that they belonged to a different class…that I did not belong to the moneyed class and um I lost interest. […] I sensed I didn't belong.

Mabel attended a Chinese Church presumably because she felt she would have a closer affinity to the congregation due to a shared cultural identity. However, the experience was, in fact, the opposite. Despite sharing some form of 'Chinese' identity, Mabel felt a stark difference between herself and other members of the Church based on class differences. Not only did Mabel communicate that she did not 'belong' to this 'moneyed class' but, as a consequence, did not belong in the Church. She resultantly 'lost interest' and ceased attending. She similarly, felt the class divides during her time at university. In Mabel's experience, the new immigrants, or individuals on student visas under the newly relaxed immigration laws, were much wealthier and of a higher social echelon than herself who is second generation Australian-born. She explained: 'there were Chinese who were ultra-rich: wore high heels, powder, diamonds and... whereas I was making my own clothes, and you sort of ... it's a class thing more than anything else' (Mabel). As such, she also felt a limited affinity to this group and socialised with them sparingly despite her position as the President of the Chinese Student Association.

Figure 6.1 Patricia Yip and her school friends, Sydney, New South Wales, c. 1955. In this photograph Patricia (second from left) poses with her school friends who she identified as Greek, Jewish and Taiwanese in ancestry. Despite the multiculturalism of her school friends, Patricia recalled the overwhelming sense of difference growing up within a predominantly 'White' society.

Source: Patricia Yip (private collection).

In my discussion of postcolonial feminist perspectives in Chapter 2, I highlighted recent attempts to reject assumptions that non-White women or 'Third world women' are 'monolithic others'. Like many postcolonial feminists, I have acknowledged throughout *Intersectional Lives* the diversity within the Chinese Australian female population to indicate that they were not a singular homogenous group, but a cohort made up of individuals with unique experiences and perspectives. Patricia and Mabel's recollections of their classed positions are thus particularly pertinent as they indicate a self-perception of this diversity within their communities. In fact, it highlights that there was not one Chinese community—*The* Chinese community—but a multitude of Chinese communities. As many researchers have previously noted, throughout their history in Australia, Chinese immigrants and their descendants have formed/maintained social groupings according to village, clan, kinship ties and dialect groups (see, for example, Teo 1971; Choi 1975; Williams 1999; and Bowen 2011). Mabel's recollection also indicates that social groupings were formed on the basis of class and thus extends Bagnall's (2006) finding that '[t]he [nineteenth century]

Chinese communities in NSW were made up of members of different social groupings or 'classes', divided along lines of wealth, education and social status according to both Chinese and western paradigms' (p: 120). By highlighting class divisions in the White Australia Policy context, Patricia and Mabel illustrated that these divisions carried on well into the twentieth century and informed the complexity of belonging and not belonging in the White Australia context.

The limits of assimilation

In examining her own experiences of identity construction and negotiation as a person of Chinese descent, born in Indonesia, raised in the Netherlands and now living in Australia, Ang (2001) noted:

> Even the most westernized non-Western subject can never become truly, authentically Western. The traces of Asianness cannot be erased completely from the westernized Asian: we will always be 'almost the same but not quite', because we are 'not White'.
>
> (p: 9 citing Bhabha 1994: 89)

Through this reflection, Ang points to the limits of assimilation and belonging for 'non-Western' individuals in 'Western' societies such as Australia. In my discussions with Chinese Australian women regarding their sense of 'Australianness', 'Chineseness' and their interactions with their White Australian communities, similar issues regarding assimilation, and more particularly, the limits of assimilation, arose. According to the interviews, during the White Australia Policy era, participants never fully assimilated—their Chineseness, whether self-ascribed or externally imposed, was never included within notions of 'Australianness'. No matter how 'Australian' participants felt, their Chineseness was brought to the fore—whether on an everyday basis in the home or at school, or on random occasions in public places and spaces.

Inherent in many Chinese Australian women's memories of 'difference' was a desire to be like the White majority—to have their difference/Chineseness erased or negated—or, in Hage's (1998: 52) terms, many participants '[strove] to accumulate nationality'. This desire to experience a sense of belonging to the national identity was communicated by Doreen:

> Subconsciously the Chinese bit is there, but I was rebelling against it. I wanted to be like the majority and their liberal, more freer lifestyle than the life I was living to my mother's strict values and constraints on life.

Similar to my discussion of women's roles in Chinese Australian families (in Chapter 4), in this recollection, Doreen draws upon Orientalist constructions of Eastern and Western values to explain her desire to abandon her 'Chineseness'. In her view, the 'Chinese' culture of 'strict values and constraints on life' was dichotomously opposed to the 'liberal, more freer lifestyle' of the White majority.

Doreen also explained, however, that her Chinese identity was immutable: 'It's just through osmosis or something. It's in your blood...the Chinese cultural background', and thus she believed she could never, in fact, truly assimilate and 'be like the majority'. Ina's recollections also provide a pertinent example of an inability to be truly 'Australian'. Ina explained in her interview that she 'melted in with the majority of Australians' and 'became Australianised' in her childhood years. Thus despite being Australian-born and having rights to Australian nationality, her Chineseness placed her beyond the boundaries of the racialised nation that was White Australia. She felt she had to 'become' Australian, despite her rights of birth.

In his discussion of the construction of nation and national belonging, Hage (1998: 52) asserted that those who strive to accumulate nationality:

> [...] recognise themselves as more national than some people and less national than others. They are also recognised by others in a similar fashion. This is made quite clear when examining the everyday language referring to national belonging in Australia and which constantly negates an either/or conception of Australianness. People talk of others as being 'almost Australians'. One migrant can refer to another migrant saying: 'He is more of an Aussie than I am.' Such a 'more or less' qualification is also used between non-migrant Australians.

Interestingly, later in her interview, Ina made such exact claims, comparing herself to a Chinese Australian friend and explaining: 'She's much more Australianised than I am, although I think I am about the limit, too'. Not only did Ina's claim highlight assumptions that individuals can obtain 'more or less' nationality, but follows arguments made by Bauman (1991) and Ang (2001) that for those beyond the national imaginary, nationality/belonging/assimilation can never fully be obtained/achieved. Ina believed she was at the 'limit'—she could never truly be Australian, her sense of difference remained. This was pertinently highlighted in one of her recollections from school:

> Yeah, and I knew I was different in one way, because my mother's name... they used to say it...they call out every now and again at the beginning of the year...you have to call out your mother's name and they'd write it down. And my mother's name was Kwan and I used to die when I had to say that. I used to absolutely die. And I can remember, you know, almost wetting my pants when my turn came to call out my mother's name. I can remember that. [...] I wanted to fade into the background. I didn't want to be different, especially with a mother named Kwan.

In this instance, Ina's sense of difference was not determined by what other Australians said or did, but her assumptions of her own 'foreignness'. Despite her Australian birth, Ina believed her mother's name located her beyond the national imaginary of 'true' Australia.

In fact, for some women assimilation was simply not an option or sought after. They did not wish to abandon their Chineseness but claimed it as an inherent and essential part of self. Mary, for example, who migrated to Australia with her sister at age 14, explained the links between her identity, assimilation and her maintenance of Chinese culture:

> I'm probably more assimilated in this society than in Hong Kong society, but on the other hand you cannot cut off that bit of you that's always been, you know that Chinese person. Since fourteen I came here, you know. I could read [Chinese], I could write [Chinese]. I still read [Chinese], I still write [Chinese]. I teach Chinese calligraphy for God's sake. So that part of my culture is very deep. I cook Chinese food, I eat Chinese food. So, you know this thing of *being* Australian or *being* Chinese is not something you can just say you are one thing or another.

For Mary, her Chinese cultural maintenance is clearly linked to her continued sense of Chinese identity. Thus, like many of the other participants, Mary's voice suggests that Chinese heritage is inextricably linked to identity no matter how 'Australian' participants are or feel. Her words are also an exemplar of the complex identity negotiations participants experienced in the White Australia period and in contemporary multicultural Australia.

It must be kept in mind, however, that the conflicting identities Ang (2001), and Hage (1998) and the interview participants refer to are based on constructed ideals of what it means to be 'Western'/Australian or 'Asian'/Chinese. Interview participants recalled an inability to fully assimilate and truly conform to an idealised 'Australian' identity that was defined by White Australia discourse and racial ideology. This is not to say, however, that they were not 'Australian'. A postcolonial approach provides a means to understand the complex and shifting affinities to various national/cultural/racial identities experienced by participants as manifestations of 'Australianness' which often involved being both 'Chinese' and 'Australian'. Viewed through this lens, these alternative ways of being 'Australian' challenged assumptions of the dichotomous and mutually exclusive categories of national/cultural identity.

Experiences of inclusion

It is important to note that not all interview participants reported experiences of racial discrimination or prejudice and subsequent feelings of being an 'outsider'. Many participants asserted that they were 'lucky' that they did not encounter much racial discrimination and explained that they were, in fact, largely unaware that the White Australia Policy even existed. For example, Edith explained that she 'didn't come across a lot of prejudice' while Nancy similarly recalled 'I was quite happy about being there. I can't recall any racial prejudice'. Interestingly, Mary, who reported one experience of racial prejudice in which she was told to 'go back' where she belonged (discussed earlier in this chapter) was sympathetic to those White Australians that often made her feel like an outsider, she explained:

I have to say generally and personally I haven't really felt any discrimination because of my colour, or race. You know, you can't hold it against people who talk to you slowly because they think you can't speak English.

Daphne, who experienced the legislative restrictions on Chinese female immigration first hand (see Chapter 3), was similarly positive, she asserted:

Otherwise, basically I would have to say I really don't have too many negative experiences here in Australia. The funny thing is, talking about negative experiences, it's almost the opposite.

Assertions such as these were not uncommon and, such as in Daphne's case, were often communicated to me even after an experience of racial discrimination or prejudice had been recalled. Similarly, many participants also emphasised that their gendered or classed identities did not mark them as outsiders. Therefore, I will now shift my focus to participants' recollections of positive interactions with White and Chinese Australians and argue that these interactions often fostered feelings of inclusion.

Crossing cultural, 'racial' and often economic boundaries, many participants formed strong and lasting friendships with non-Chinese Australians within the White Australia context. As I discussed in the previous chapter (Chapter 5), some interview participants often felt extremely isolated from other Chinese Australians in their childhood and adolescent years due to their residence in outlying suburbs of Sydney or rural towns in New South Wales and Queensland. In that way, they had 'no choice' but to socialise and interact with the broader White Australian community. Despite holding on to many Chinese cultural practices and traditions within the childhood home (often as a consequence of their parents' wishes) and regardless of experiences of discrimination and prejudice encountered within their schools and broader communities, the majority of interview participants recalled the ease with which they integrated into Australian social life via friendship groups. For example, when I asked Mei-Lin if any of her White Australian school friends cared that she was Chinese, she responded by saying:

No, no! I mean my best friend was a blonde, blue eyed child who lived three doors down from where we lived and my next door neighbours were Irish Australians from Tasmania. Mum was a member of the Country Women's Association. They managed...they seemed to move into the community quite easily. They both found it quite easy to mix into the community and it was the same when we moved to Ballina.

In this response, Mei-Lin suggested that she and her family actively sought to participate in community life and were successful, to a great extent, in their attempt to assimilate. As Mei-Lin went on to explain: 'My parents obviously decided to raise us to be part of...being open and friendly and therefore people accept you as such [...] and therefore our path was, as I said, quite easy'. From Mei-Lin's perspective, assimilation and the forming of social bonds with White Australians facilitated an 'easy' life relatively free of the burdens of otherness (Figure 6.2).

In a similar way, other interview participants also recalled fond memories of the childhood friendships they formed through interactions with non-Chinese Australians in schools and neighbourhoods. Marina, for example, explained the protection and support she received from her 'Australian' friend against school-yard bullies, while Ina recalled playing with her 'the Australian girls up the road' and correlated this socialisation with a lack of discrimination. For Doreen, the family home of her childhood friend, Nancy, provided her with a safe and supportive environment, a 'home away from home' in which she felt appreciated and free from the constraints of her Chinese upbringing. She explained:

> I spent as little time I possibly could at home. I was always up at Nancy's or whatever. [...] I didn't like the life and if I went back home [... my brothers] always got prime everything and when I went up to Nancy's her mother made me feel like a person. So that was why I spent a lot of time at Nancy's home.

It is interesting to note that Doreen's desire to escape her family situation was rooted in the patriarchal ideologies that shaped her home life. She indicated that the patriarchal system in her home was oppressive and that she experienced more positive interactions in Nancy's home where the building of her own sense of self-worth was facilitated by Nancy's mother. Doreen's experience of family patriarchy therefore pushed her towards cross-cultural interactions. Thus although Marina, Ina and Doreen all recalled instances in which their 'Chineseness' led to a real sense of difference from the White majority, it was these positive 'everyday' interactions with White Australians, particularly in their childhood, which informed their overarching sense of belonging and inclusion in the White Australia Policy context.

Figure 6.2 Doreen Cheong celebrating a birthday, Ballina, New South Wales, c. 1955. In this photograph, Mei-Lin (first on left) poses with her classmates at a birthday party in her home town. Mei-Lin explained that her family were the only Chinese Australians in Ballina at that time. She therefore socialised with her White Australian peers and explained that her family easily integrated into the broader White community.

Source: Mei-Lin Yum (private collection).

Interview participants' recollections of childhood friendships often drew upon the assumption that children were ignorant or oblivious to the 'colour divide'. As Eileen explained:

> But the neighbours, they were all fantastic. Everyone went through hardships in the early days. We played with the kids in the street. We didn't know what colour was. It was only the politics and all that later on.

Obliviousness to 'racial' difference among children was also suggested in Stella's memory of playing the childhood game 'Picking Johnny's Cabbages':

> Well the amusing thing was that they played little racist games, like there was a game called 'Picking Johnny's Cabbages'. That was the market gardener I presume, Johnny. And you'd all run over and somebody was Johnny and you'd all pretend to pick his… steal his vegetables. And he'd chase you. It was a good game and we all joined in. I never realised or never cared either. We all played together. There were little incidents there of racism but on the whole, with about fourteen different ethnic groups on the Island, there's very little disharmony.

Rooted in the racial stereotypes associated with the Chinese market gardener, the game, in Stella's view, did not draw attention to her Chineseness or serve to mark her as different and 'other', but was an innocent way in which she interacted with her school friends. For them, the racial undertones were irrelevant.

In the previous section, I highlighted the ways in which interview participants such as Doreen and Patricia felt a real class consciousness rooted in their family's low socio-economic position. In some instances, rather than being outcast because of this 'difference', help was often at hand with White Australian neighbours and friends providing goods, assistance such as language skills, and cultural knowledge when it was needed. Doreen in particular noted these types of positive interactions in her childhood years, she recalled:

> [In] those early years we were very grateful for a lot of our Australian neighbours around us because they must've thought, 'Oh these poor Chinese, living on the smell of an old rag' and they gave us clothes and hats and there were always hand-me-downs and people making things and doing things for us which would've been very, very difficult. I know one man who lived in the next street was a handy man but he also helped fill out some of the bureaucratic papers that we had to fill out and all that kind of thing.

This recollection, however, highlights the complexity of such everyday interactions. Although non-Chinese friends and neighbours were well-intentioned and

Doreen was 'very grateful' for the acts of charity and assistance they provided, Doreen perceived that she was seen as an outsider—not only poor but also 'Chinese'. In this way, help and assistance seemed to ingrain or emphasise the sense of difference between her and wealthier White Australians.

Beyond acts of charity in neighbourhood communities, help and assistance were most commonly highlighted in participants' recollections of school. In contrast to some participants' recollections of school-yard taunting and bullying, many participants remembered the special treatment they received from teachers to help them adjust to the school environment. Australian-born Kaylin explained: '[...] the teacher's treated us a little special because we were different'. Marina, who was also Australian-born, recalled a similar experience: 'They were all very nice, the teachers. [..] I can't remember any discrimination; maybe more favouritism than discrimination'. Positive student–teacher interactions at school were similarly experienced by migrant students such as Helen and Patricia:

> The nuns were so nice, because she knew I didn't understand. She gave me extra tuition.
>
> (Helen)

> I don't remember much about the teachers, but there was one very nice teacher called Miss Burke and she was my third class teacher. I think she was very sympathetic towards the new migrants. I don't know, I just feel she was really nice and helpful.
>
> (Patricia)

These recollections provide an alternative account of the school experience of Chinese Australian students during the White Australia Policy era as documented by Tan (2003). In her exploration of growing up 'Chinese' in White Australia, Tan argued that educational practices and actions of teachers reflected assimilationist policies and were perceived negatively by Chinese Australians. These actions included placing pressure on children to 'shed "Old World traits" and become as "Australian" (read: White) as possible' (Tan 2003: 104) by 'becom[ing] proficient in English, speak[ing] with an Australian accent, and adopt[ing] White Australian behavioural norms, customs and practices' (Tan 2003: 104; see also Wilton and Bosworth 1984; and Wilton 1996). In fact, many of Tan's respondents recalled being punished for speaking Chinese in the classroom or failing to understand teachers' instructions (Tan 2003: 104). In contrast, the interview extracts above suggest a more positive view of the 'special treatment' participants received from their teachers, including help learning the language and meeting school expectations. From the perspective of interview participants, these actions were not rooted in a disapproval of 'Chineseness' but were intended to provide them with the tools to achieve cultural competency and success in their education.

Interviews with Chinese Australian women therefore provide a more complex and nuanced insight into the interactions between these women and non-Chinese Australians. While cross-cultural contacts were often based on discriminatory interactions or racist incidents, more often than not, interview participants had positive encounters with non-Chinese Australians and formed intimate bonds of friendship and social support. This suggests that 'White Australian' ideologies were not uniformly inscribed into the consciousness of Australians. The complex nature of interview participants' relationships with non-Chinese Australians is further highlighted as interviews indicate that cross-cultural interactions—particularly being treated especially well because of being 'Chinese'—facilitated a sense of belonging for some and exacerbated a sense of difference in others.

Conclusion

In this chapter, I have focused on the public presence of Chinese Australian women in the White Australia Policy era to explore their experiences in contexts beyond the home—such as in schools, workplaces, local neighbourhoods and communities—and, more particularly, examined their interactions with other Australians in these spaces. By examining the recollections of Chinese Australian women, I have argued that racial discriminatory practices or prejudice were not uniformly experienced by Chinese Australian females. Racialisation took many forms in many different spaces—ranging from name-calling at school, being asked 'where are you from?' by strangers on public transport, to institutionalised discrimination in education systems and places of employment. In these instances, participants' 'Chineseness' was defined by White Australians and marked participants as 'Other' or beyond the scope of national belonging. However, the analysis of Chinese Australian women's recollections also indicated that these women were not submissive 'victims' of racialisation; they negotiated and asserted their own identities in these racialised contexts and often challenged assumptions of the mutually exclusive nature of 'Chinese' and 'Australian' identities. Furthermore, Chinese Australian women's everyday experiences in spaces beyond the home were not always marked by racial discrimination or prejudice. Some participants formed strong and lasting friendships with White Australian peers or were supported and encouraged by White Australian neighbours and teachers. Through these relationships, Chinese Australian females obtained a sense of inclusion and belonging.

This chapter is therefore positioned as an important contribution to White Australia/Chinese Australian research. By utilising a postcolonial feminist approach that privileges research 'from below' methods, I have uncovered a diverse range 'public' experiences as told from Chinese Australian women's own point of view. In this way, I have not only reiterated the presence of Chinese Australian women in the nation throughout the White Australia policy (and in public contexts more specifically), but also made visible their often-overlooked experiences and personal perspectives. The analysis of the nuanced accounts of personal experiences presented in this chapter therefore contributes to the correction of existing research that has focused on the experiences of Chinese Australian men in the public domain and essentialised the experiences of Chinese Australians as racialised subjects.

By providing an opportunity for Chinese Australian women to speak of their interactions, I have provided insight into the ways in which discrimination and prejudice impacted their lives. From ostracising experiences in shops, to gendered and racialised discrimination in the education system and workplace, interview participants 'race' and gender had real impacts on their interactions with non-Chinese Australians and in some instances, significant effects on their life opportunities (particularly in regards to the education system and employment opportunities). These findings not only contribute to existing male-dominated understandings of the Chinese Australian experience in the White Australia Policy era but also to understandings of the way in which national discourses in the period constructed belonging.

Note

1 A national survey of Asian Australians conducted by Kamp et al. (2021: 15) in 2020 found that the most common setting Asian Australians experienced racism were public settings such as in shops and on the street.

References

Anderson, B. 1983, *Imagined Communities: Reflections on the Origin and Spread of Nationalism*, Verso, New York.

Ang, I. 2001, *On Not Speaking Chinese: Living Between Asia and the West*, Routledge, London.

Bagnall, K. 2006, *Golden Shadows on a White Land: An Exploration of the Lives of White Women Who Partnered Chinese Men and their Children in Southern Australia, 1855–1915*, unpublished PhD thesis, University of Sydney, Sydney.

Bauman, Z. 1991, *Modernity and Ambivalence*, Polity Press, Cambridge.

Bhabha, H. 1994, *The Location of Culture*, Routledge, London.

Blair, K., Dunn, K., Kamp, A., Alam, O. 2017, *Challenging Racism Project 2015–2016 National Survey*, Western Sydney University, Sydney.

Bowen, A. 2011, 'The merchants: Chinese social organisation in colonial Australia', *Australian Historical Studies*, vol. 42, no. 1, pp. 25–44.

Cerulo, K. A. "Identity Construction: New Issues, New Directions." *Annual Review of Sociology*, vol. 23, no. 1, 1997, pp. 385–409.

Choi, C. 1975, *Chinese Migration and Settlement in Australia*, Sydney University Press, Sydney.

Cornell, S. and Hartmann, D. 1998, *Ethnicity and Race: Making Identities in a Changing World*, Pine Forge Press, London.

Dunn, K. M. 2005, 'Repetitive and troubling discourses of nationalism in the local politics of mosque development in Sydney, Australia', *Environment and Planning D: Society and Space*, vol. 23, pp. 29–50.

Dunn, K. M. and Ip, D. 2008, 'Putting transnationalism in context: Comparing Hong Kong Chinese-Australians in Sydney and Brisbane', *Australian Geographer*, vol. 39, no. 1, pp. 81–98.

Elder, C. 2003, 'Invaders, illegals and aliens: Imagining exclusion in a white Australia', *Law Text Culture*, vol. 7, pp. 221–250.

Elder, C. 2005, 'Immigration history', in M. Lyons and P. Russell (eds.) *Australia's History: Themes and Debates*, UNSW Press, Sydney, pp. 98–115.

Fitzgerald, J. 2007, *Big White Lie: Chinese Australians in White Australia*, University of New South Wales Press, Sydney.

Fitzgerald, S. 1997, *Red Tape, Gold Scissors: The Story of Sydney's Chinese*, State Library of NSW Press, Sydney.

Gilroy, P. 1987, *'There Ain't No Black on the Union Jack': The Cultural Politics of Race and Nation*, Routledge, London.

Hage, G. 1998, *White Nation: Fantasies of White Supremacy in a Multicultural Society*, Pluto Press, Sydney.

Jackman, M. 1994, *The Velvet Glove of Paternalism and Conflict in Gender, Class and Race Relations*, University of California, Berkley, CA.

Jackson, P. and Penrose, J. 1993, 'Placing "race" and nation', in P. Jackson and J. Penrose (eds.) *Constructions of Race, Place and Nation*, UCL Press, London, pp. 1–26.

Jeung, R., Yellow Horse, A., Popovic, T. and Lim, R. 2021, *Stop AAPI Hate National Report*, Stop AAPI Hate coalition (online).

Kamp, A. 2010, 'Formative geographies of belonging in White Australia: Constructing the national self and other in parliamentary debate, 1901', *Geographical Research*, vol. 48, pp. 411–426.

Kamp. A., Denson, N., Atie, R., Dunn, K., Sharples, R., Vergani, M., Walton, J., and Sisko, S. 2021, *Asian Australians' Experiences of Racism During the COVID-19 Pandemic*, Centre of Resilient and Inclusive Societies, Deakin University, Melbourne.

Noble, G. (ed.) 2009, *Lines in the Sand: The Cronulla Riots, Multiculturalism and National Belonging*, Institute of Criminology Press, Sydney.

Palfreeman, A. 1967, *The Administration of the White Australia Policy*, Melbourne University Press, Melbourne.

Penrose, J. 1993, 'Reification in the name of change: The impact of nationalism on social constructions of nation, people and place in Scotland and the United Kingdom', in P. Jackson and J. Penrose (eds.) *Constructions of Race, Place and Nation*, UCL Press, London, pp. 27–49.

Pratt, G. and Hanson, S. 1994, 'Geography and the construction of difference', *Gender, Place & Culture: A Journal of Feminist Geography*, vol. 1, no. 1, pp. 5–29.

Shaw, W. S. 2006, 'Decolonizing geographies of whiteness', *Antipode*, vol. 38, pp. 851–869.

Tan, C. 2003, 'Living with 'difference': Growing up 'Chinese' in white Australia', *Journal of Australian Studies*, vol. 77, pp. 101–108, 195–197.

Tan, C. 2006, '"The tyranny of appearance": Chinese Australian identities and the politics of difference', *Journal of Intercultural Studies*, vol. 27, no. 1–2, pp. 65–82.

Teo, S. E. 1971, 'A preliminary study of the Chinese community in Sydney: A basis for the study of social change', *Australian Geographer*, vol. 11, pp. 579–592.

Williams, M. 1999, *Chinese Settlement in NSW: A Thematic History/A Report For the NSW Heritage Office*, NSW Heritage Office, Parramatta.

Wilton, J. 1996, *Chinese Voices, Australian Lives*, unpublished PhD thesis, University of New England, Armidale.

Wilton, J. and Bosworth, R. 1984, *Old Worlds and New Australia: The Post War Migrant Experience*, Penguin, Ringwood.

Yamamoto, T. 1999, *Masking Selves, Making Subjects*, University of California Press, Berkley, CA.

Yang, W. 1994, 'I ask myself, am I Chinese?' *Art and Asia Pacific*, April, pp. 89–95.

Yu, H. 2001, *Thinking Orientals*, Oxford University Press, New York.

7 Conclusions and looking forward

In the introduction to this book, I argued that researchers have perpetuated notions that Chinese Australian females were absent from the Australian nation for much of the twentieth century or have overlooked their experiences and contributions. Thus, a 'gender-blind' history has been maintained. Given that Chinese Australian females *were* present in the Australian nation throughout the White Australia Policy era (as I have demonstrated throughout this book), these androcentric perspectives have had detrimental impacts on scholarly (and broader public) understandings of the social, economic and political participation of Chinese in White Australia and their contribution to Australian nation-building. In broad terms, by overlooking or erasing the presence, activities and contributions of Chinese Australian females, the nuances and complexities of Chinese migration to Australia and settlement experiences have only been partially uncovered. As I will summarise below, this has had flow-on effects on our understanding of the White Australia context, 'Australian' family life and intra-family relations, female mobility, migrant family economies, Australian cultural pluralism and broader understandings of Chinese diaspora experience.

Intersectional Lives has therefore been driven by a postcolonial feminist objective to make visible the presence and intersectional experiences of Chinese Australian females in the White Australia Policy era. By focusing on Chinese Australian females that were present in the nation, I have presented a more inclusive (but by no means complete) account of Chinese Australian experience. This project has built upon the small body of research that has begun to 'put women into' histories of Chinese in Australia, including the work of Giese (1997), Tan (2003), Couchman (2004), Khoo and Noonan (2011), Martínez (2011, 2021) and Gassin (2021). However, by uniquely placing Chinese Australian females at the centre of research and providing a space in which their voices can be heard in academic scholarship, I have more stridently answered calls to undertake inclusive research that gives voice to Chinese Australians (Cushman 1984; and Chan 1995, 2001) and correct 'gender-blind' Chinese Australian scholarship (Loh 1986; Chan 1995; Bagnall 2006; and Kamp 2013).

Closely aligned to these 'political' objectives was the more scholarly intention to utilise geographical and postcolonial feminist approaches to reveal the complexity and diversity of Chinese Australian female experience in White Australia. I aimed to demonstrate that intersectional analyses that focus on

DOI: 10.4324/9781003131335-7

females' multiple identities, experiences and marginalisations within various power structures and relations are crucial to redressing androcentric and ethnocentric research approaches that have dominated Western research traditions. I also sought to utilise intersectional approaches to ensure that 'monolithic' representations of 'oppressed' and 'victimised' Confucian women were not perpetuated. From a historical–geographical perspective, understanding how the diverse and intersectional experiences of Chinese Australian females shifted across time, place and space was a priority.

To guide these broad political and scholarly intentions, *Intersectional Lives* centred on four themes which pertained to: the presence, roles and contributions of Chinese Australian females; the interactions between Chinese Australian females and other Australians; the range of identities constructed and negotiated by Chinese Australian females; and the power structures that influenced the lived realities of Chinese Australian females in the White Australia context. The result is a postcolonial feminist historical geography of Chinese Australian females' experiences in White Australia.

The adoption of geographical and postcolonial feminist approaches not only played a significant role in crafting the overall intention of this book, but also the methods I employed for the data collection and analysis that is presented therein. The first method I employed was the collection and analysis of historical census data for the years between 1911 and 1966. These quantitative data were used to uncover population numbers and various demographic characteristics of the female Chinese Australian population (that is, those females racially classified as 'Chinese') throughout the White Australia Policy era. This included birthplace, length of residence, state of residence, occupation and literacy. Census data are 'official' data of the State and the racial classifications were utilised by the Australian government to identify, count and categorise undesirable 'Other' groups as divergent from the narrow White national identity. Significantly, however, I did not utilise the racial census data for means of exclusion; rather, through a postcolonial feminist approach, they were used as a means to *include* a 'subaltern' group. This has importantly demonstrated the way in which colonialist (and patriarchal) tools can be used for postcolonial (and feminist) purposes.

The second method I employed was the collection and analysis of first-hand accounts of the experiences of Chinese Australian females via in-depth interviews with 19 women. In keeping with my postcolonial feminist epistemology, the interview method privileged the voices of females and allowed some of those individuals that were documented in the censuses to be fleshed out *as people* with lived experiences, memories and identities. By combining the interview method (and collection of personal documents) with an analysis of census data, I was able to re-examine and revise the largely androcentric accounts of population numbers and Chinese Australian migrations and presence. I was also able to focus on interview participants' everyday contexts, acknowledge and assess their varied experiences and multiple identifications over time and place, and examine the various power structures that impacted their lives. This research method enabled me to fill some of the gaps in Chinese Australian scholarship and was used for its postcolonial capabilities, that is, a means to move away from patriarchal and colonialist research traditions in historical geography and the broader social sciences.

Key findings and contributions

The research presented in this book has not only emphasised the presence of Chinese Australian females throughout the White Australia Policy era, but also revealed a clearer picture of their diversity and intersectional experiences. Chinese Australian female lives and experiences were not 'monolithic', but shaped by various identities, subjectivities and manifestations of patriarchy, racism and other structures of power.

Presence and mobility

Accounts of Chinese females in existing literature regarding nineteenth- and twentieth-century Chinese immigration and settlement to Australia are generally rooted in patriarchal and Orientalist discourses that assert their position as 'victims'—victims of traditional Confucian culture and/or victims of restrictive immigration legislation. In this way, they have often been constructed as absent or dependent wives. My research has challenged such limited understandings of female Chinese experience. For example, my analysis of historical census data indicated that when both 'full' and 'mixed' Chinese Australian females (as defined by the census statisticians) were included in population counts, the number of Chinese Australian females was much larger than has been recorded in existing literature (such as in Choi 1975; Yong 1977; Cushman 1984; and Fitzgerald 2007). This quantitative evidence (alongside my analysis of settlement histories collected via the interview process) also indicated that this presence was not momentary or sporadic, but long and enduring. Some women had migrated to Australia prior to the 1890s and had resided in the country for over 20 years while other females (Australian-born interview participants) recalled the multiple generations of women in their family who had resided in Australia before them. Unlike the population of Chinese Australian males which declined after the passing of the national *Immigration Restriction Act 1901*, the number of Chinese Australian women steadily increased throughout the period. This increasing population of women was (unevenly) spread across all the Australian states and territories and in urban and rural locations. Like their male and non-Chinese counterparts, Chinese Australian females were predominantly located in the eastern states (Queensland, New South Wales and Victoria) and in urban centres throughout the period.

The demographic diversity of Chinese Australian women was also highlighted in the wide array of countries of origin. While the majority of Chinese Australian women resident in Australia in the census years between 1911 and 1966 were Australian-born, many other international birthplaces across Australasia, Europe, Asia, the Americas, Africa and Polynesia were represented. These birthplace statistics, alongside migration data and firsthand accounts of migration experiences, indicated that females were internationally (and intra-nationally) mobile—a phenomenon that challenges general assumptions of female immobility in global migration patterns and understandings of female Chinese migration to Australia. Females migrated as dependent children, students and wives (but not always following the traditional or expected route of passive or dependant housewives following their migrant husbands). Some Chinese women occupied empowered positions in the migration process—being able to negotiate and work around immigration restrictions rather than being passive victims of discriminatory policy (as is most commonly perpetuated in existing research). By combining birthplace information recorded in the censuses and interview participants' accounts of their migration history, I uncovered various reasons and motivations for this movement across international boundaries—including study, economic opportunity and marriage—and highlighted the diversity of the migration experience.

These findings regarding female mobility and migration not only make substantial contributions to the area of Chinese Australian research, but broader understandings of overseas Chinese migrations and settlement experiences. Alongside the work of Ip (1990, 1995, 2002) in New Zealand and Yung (1998) and Ling (1993, 2000) in North America, this project has demonstrated that international movements of Chinese in the nineteenth- and twentieth centuries were not simply movements of men—Chinese women were also migrating across borders, challenging Confucian female ideals and contributing to the development of diasporic communities. The identification of such movement also contributes to broader migration research which, as I indicated in Chapters 1 and 3, has typically constructed the migrant as male.

Homes and family responsibilities

By prioritising the voices and perspectives of Chinese Australian females and focusing on the spaces in which they lived and worked, *Intersectional Lives* has shifted focus away from the experiences of Chinese Australian males in their publicly visible political and economic roles. Instead, I have explored female experiences as daughters, wives and mothers in the intimate contexts of home and family life. This enabled a further correction of previous constructions of absent wives and the 'lack of family life' within Chinese Australian 'bachelor' communities. The complex and diverse nature of interview participants' (and their mothers') experiences and contributions to the family unit/family economy were uncovered. Many of the females featured in this book contributed to the running of households in ways that conformed to traditional gender expectations—cooking, cleaning and caring for children. Some also helped in the running of family businesses—either 'out the back' (as defined by Song 1995) in ways that conformed to their traditional gender roles, or in more visible positions where they took on greater responsibility than traditional gender roles would have prescribed. Some Chinese Australian females (more frequently interview participants rather than their mothers) participated in paid employment beyond family businesses and thus made important financial contributions to their families. Although small in number, the presence and contributions of these females challenge existing notions of the subordinate and dependant woman confined to the domestic sphere in accordance with patriarchal ideals.

The 'double burden' of maintaining the home as well as working in family businesses or in alternative employment was a reflection of women's gendered, classed and cultural identities. With discriminatory policies as a backdrop, many Chinese Australian women were part of ethnic minority/migrant families that were trying to make successful lives. Within this context, women (and children) were often required to contribute to the family's financial security. Therefore, while patriarchal values maintained in families most often prescribed females be relegated to the home to conduct domestic duties, within the White Australia context this discrete division of gendered labour and space was not always possible. The visibility of these experiences begins to correct male-dominated understandings of the survival of Chinese Australian communities in twentieth-century Australia. As I have demonstrated, females were central to the survival of Chinese Australian families and communities.

The varied experiences and perspectives of Chinese Australian females also challenge traditional feminist assertions that 'the family' is universally a site of female oppression as well as Orientalist assumptions that Confucian family systems were practised uniformly by all overseas Chinese (as similarly argued by Teng 1996). Chinese Australian women were not uniformly confined by patriarchal ideals of subservience, but experienced alternative roles and subjectivities within Chinese Australian family contexts. In this way, 'the family' was not singularly a site of female oppression, but a social unit in which patriarchal systems of power were negotiated, adapted and challenged. Furthermore, despite White, Western assumptions that Chinese women are more oppressed than their Western counterparts due to the patriarchal systems aligned with Confucian ideology, I have argued that while Chinese Australian female's experiences of patriarchy cannot be conflated with broader conceptions of Western patriarchy, they were not uniquely 'Chinese'. Their experiences were individually shaped by patriarchal systems, cultural values and unique social and economic circumstances. The intersection of their ethnic/cultural identities, gender identities, class, age and generation also influenced their roles and contributions to homes and families in the White Australia Policy era. Therefore, just as scholarship has argued that there was/is no single 'traditional' Chinese family (such as Watson and Ebrey 1991; and Teng 1996), my examination has illustrated that there was also no single 'Chinese Australian' family experience in the White Australia context. The specific roles and expectations placed on female family members varied in diverse ways so that women experienced alternative subjectivities that were neither 'Chinese' nor 'Australian', but varied forms of 'Chinese Australian'.

The inability to generalise the Chinese Australian female experience according to Confucian ideals (or patriarchal ideals more generally) was further highlighted by the analysis of interview participants' recollections of their own childhood family systems and relations. In discussions of their experiences growing up, some participants explained that their upbringing was marked by gender differences in the treatment and opportunities made available to sons and daughters. In these instances, interview participants recalled being prepared for their later role as dutiful wives. For example, they were tasked with helping their mother in the domestic duties and caring for siblings; they often lacked the same educational opportunities as their brothers and were made aware that it was their role as daughters to be 'married off' in later years. Other participants (the majority) recalled that they and their siblings—both brothers and sisters—were treated equally and encouraged in their education and employment opportunities. As such, the roles and expectations placed on daughters in the Chinese Australian childhood home and family context were diverse—there was not a single or uniform experience of oppression nor was there a single or uniform freedom from that oppression in the Australian context. Interestingly, regardless of their childhood experiences, interview participants drew upon Orientalist binaries between 'Eastern' and 'Western' patriarchy to explain their family experiences. They aligned gender inequality with 'traditional' Chinese culture or Confucian ideals and female liberation/equal opportunity with 'Western'/ 'Australian' ways of thinking. That is, those who experienced gender discrimination in the home equated it with the Confucian family system, while those who experienced gender equality aligned it with an adoption of more 'Australian' ways.

Cultural maintenance and pluralism

With policies of assimilation as a backdrop and in a context of social isolation from other Chinese Australians, homes often provided important spaces in which Chinese cultural practices were maintained and cultural knowledge was passed down through the generations. Interview participants who grew up in Australia identified language, food culture and festivities/celebrations as the primary ways in which parents maintained links to 'homeland' and instilled a sense of 'Chineseness' in their children. However, there was a diversity of experience—with some families challenging 'White Australia' concepts by maintaining everyday cultural practices and more overt celebrations of Chinese identity and culture such as Chinese New Year, while in the lives of other participants, Chinese culture and identity were de-emphasised in order to conform to broader Australian society. The extent to which interview participants practised Chinese cultural traditions in their childhood was largely dependent on their parents' desire and ability to maintain aspects of traditional Chinese life.

While some fathers were pivotal in the reproduction of Chinese culture and identity in Chinese Australian homes and families, it was more commonly the case that Chinese Australian mothers were centrally involved in the maintenance of these cultural practices and the passing down of cultural knowledge to their children. Mothers' roles in the maintenance of Chinese cultural practices in homes were intrinsically linked to their responsibilities for domestic duties such as cooking, cleaning, bearing and raising children. Migrant mothers cooked Chinese meals, often spoke Chinese to their children (usually because they could not speak English themselves) and passed down other cultural knowledge as part of their caring responsibilities. Chinese Australian women, particularly in their position as mothers, could therefore be seen as 'bearers of culture' (as termed by Billson 1995); a role which was an extension of their domestic tasks and part and parcel of 'doing gender'. Women's roles as 'keepers of culture' have been noted in various immigrant and ethnic minority communities in the United States and Britain (for example Mani 1993; Beo-ku-Betts 1995; Billson 1995; Dasgupta 1998; Ebaugh and Chafetz 1999 and Le Espiritu 2001; Mohammad 2007). Ratnam (2018, 2020) has also provided important insights into the Sri Lankan Tamil Australian context. This project therefore makes significant contributions by 'adding in' the (Chinese) Aus-tralian experience.

While mothers' role in the maintenance of Chinese culture in the White Aus-tralia context is clearly linked to their position within the patriarchal family hierarchy and responsibility for the domestic sphere, a postcolonial feminist per-spective allows us to view this role as not one of oppression and victimisation, but of having some aspects of empowerment. While patriarchy was maintained in many of the participants' childhood homes, the home was also a space in which women were empowered to maintain various Chinese cultural practices, as both active agents and inadvertent cultural custodians. In this way, many of the interview participants and their mothers resisted assimilation in the White Australia context. This parallels arguments made by hooks (1991), Honig (1994), Young (1997) and Legg (2003) on the liberating potential of 'home' for women in oppressive societies.

Interactions and identity beyond the home

In Chapter 6, I shifted focus away from the domestic space of the home and family institution to examine Chinese Australian women's experiences in more public contexts and, more particularly, to examine their interactions with other Aus-tralians in these spaces. This included schools, workplaces, local neighbourhoods and communities. Interview participants asserted that their racialised 'Chinese' identities were often constructed and negotiated in public spaces and in their interactions with non-Chinese Australians. For example, participants experi-enced racialisation in the form of name-calling at school, being asked 'where are you from?' by strangers on public transport, to institutionalised discrimina-tion in education systems and places of employment. In these instances, partic-ipants' 'Chineseness' was defined by White Australians and marked participants as 'Other' or beyond the scope of national belonging. However, a postcolonial feminist analysis of Chinese Australian women's recollections also indicated that these women were not submissive 'victims' of racialisation; they negotiated and asserted their own identities in these racialised contexts and often chal-lenged assumptions of the mutually exclusive nature of 'Chinese' and 'Austral-ian' identities. The analysis also highlighted that racial discriminatory practices or prejudice were not uniformly experienced by participants. Some interview participants formed strong and lasting friendships with White Australian peers or were supported and encouraged by White Australian neighbours and teach-ers. Through these relationships, Chinese Australian women obtained a sense of inclusion and belonging.

(Re)constructing Chinese Australian historical geographies

In broad terms, *Intersectional Lives* has made important feminist and postcolonial revisions to Australia's historical geography not only by 'putting in' a group of non-White women into the historical geography of White Australia, but, more particularly, by revising dominant understandings of national belonging and exclusion in the White Australia national context. For example, despite exclusionary measures to ensure the development of a 'White' Australia, the presence and experiences of Chinese Australian women suggest the existence of cultural pluralism prior to the introduction of formal multicultural policy in the 1970s. This multiculturalism challenged White Australian national discourse which was premised on singular notions of Australianness (culturally and racially) encapsulated by Whiteness and in many ways, masculinity. I have demonstrated that such informal multiculturalism was reliant on the contributions of Chinese Australian women themselves, particularly in their roles in the domestic sphere as 'bearers of culture'. Furthermore, women's feelings of 'Australianness', alongside the presence of Australians who were racially defined as 'Chinese', indicates official and unofficial national belonging that deviated from the constructed White Australian ideal.

The contributions of *Intersectional Lives* to Australian historical geography are particularly significant given the continued disregard of women, particularly non-White women, in the sub-discipline. As I detailed in Chapter 2, the body of research that has begun to include women in Australian historical geographies continues to be small and includes work by Teather (1990, 1992), Anderson (1995), Gleeson (2001) and McKewon (2003). Australian historical geographies that focus on the lives of non-White women are even rarer (see Fincher 1997 and Ramsay 2003). Therefore, while some historical geographers in the United States and Britain have quickly answered Rose and Ogborn's (1988) call to correct the gender-blindness and patriarchal assumptions entrenched in the sub-discipline (e.g., Kay 1989, 1990, 1991; Phillips 1997; Maddrell 1998; McEwan 1998; Schuurman 1998; Morin and Berg 1999; and Domosh and Morin 2003), in Australia, feminist research output in the sub-disciplinary area continues to be marginal (Gorman-Murray et al. 2018). My (re)construction of the historical geography of Chinese Australian women in the White Australia policy era is therefore positioned as an important postcolonial feminist contribution to Australian historical geography.

By uncovering and analysing the experiences of Chinese Australian women in the White Australia period, my research not only provides insight into past understandings of Australian identity, belonging and exclusion but also facilitates a clearer understanding of the contemporary Australian context and contributes to current thematic concerns within postcolonial geographies regarding identity and difference across space and place. As I detailed in Chapter 2, links between past and present are central to *Intersectional Lives* as many females at the centre of this project are still living and those that participated as interview participants provided the qualitative data used to reconstruct the past geographies. These females have also left legacies in contemporary Australia, be they cultural, economic/business, or familial. Thus, in understanding the past experiences of Chinese Australian females we are able to more fully understand and address their and their descendants' contemporary situation. In this way, *Intersectional Lives* contributes to broader debates on national identity, diversity, cohesion, multiculturalism and Australia's place within the Asian region. Given that the recognition of past injustices to Chinese Australians has been overlooked (Han 2011; and Lowe Kelley 2011), as I also outlined in Chapter 2, *Intersectional Lives* contributes to work on the politics of recognition which assert that acknowledging and recognising previously silenced histories—such as Indigenous histories in Australia—can help processes of reconciliation (Haebich 2000). In these ways the research I have presented highlights the important role of historical–geographical research in the contemporary Australian context.

Directions for future research

Given the relatively small amount of research that has focused on the lived experiences of Chinese Australian females throughout the White Australia Policy era, avenues for future research are numerous. There are, however, some specific recommendations for future research that have arisen from this research project in particular. For example, the census data collected and analysed for this project were limited to what is publicly available, that is, the census records for the years between 1911 and 1966. As I detailed in Chapter 2, racial data from the 1971 census have not been made publicly available via published census reports, however, it does exist. While it would take some work to get hold of these data (evidenced in my hours spent on the phone to the Australian Bureau of Statistics attempting to obtain the information) and collate it, it would be well worth the effort in order to complete the dataset. Similarly, collating the state records for the first national census in 1901 would also provide a more complete picture. Beyond census data, other official information regarding Chinese Australian females does exist. This includes migration records, birth certificates, death certificates, marriage certificates and Certificates of Exemption from the Dictation Test. Other researchers have begun to utilise these sources but there is still much work to be done.

This research project has shed light on the mobility and migration of Chinese females to Australia in the nineteenth- and twentieth centuries, however, many questions regarding their movement remain unanswered. In Chapter 3, I posed the following questions: did entries into the nation reflect temporary settlements, did the experiences of Chinese females who migrated to Australia follow dominant understandings of passive wives following their husbands abroad, or did their movements challenge patriarchal power structures within Chinese family systems? I was able to answer these questions in regards to the lives of the 19 interview participants and their female family members; however, an examination of more female lives and experiences would provide further insight.

My examination of Chinese Australian female experiences of cultural maintenance in the White Australia Policy era in Chapter 7 was confined to experiences which took place in the family and home context. While this focus was a reflection of the experiences primarily recalled by participants, I did note that some interview participants also recalled maintaining 'traditional' Chinese cultural practices in contexts beyond the home and family such as in university student groups and community organisations such as the Chinese Women's Association. Given the confines of this book, I was unable to provide a detailed analysis of these experiences. Further research into this aspect of cultural maintenance would be extremely beneficial for making visible Chinese Australian females more 'public' demonstrations of their 'Chineseness' and their additional contributions to the development of Australia's Chinese cultural landscape and multiculturalism more broadly. Such research would also aid in further complicating dominant understandings of national identity and culture in the White Australia Policy era.

Throughout the chapters of this book, I frequently discussed notions of Confucian patriarchy and 'traditional' Chinese culture which participants believed shaped their experiences in White Australia—in terms of gendered roles and maintenance of Chinese cultural practices. I was not able to investigate the origins of participants' perceptions of this cultural divide between China and Australia which, for the most part, constructed Chinese culture are backward and oppressive to females, and Australian culture as more progressive and liberating. While these perceptions among many of the interview participants were Orientalist in nature, it is difficult to assume that Western Orientalism had such a strong foothold in Chinese Australian families and influenced Chinese Australian female attitudes. Questions therefore arise regarding where/how Chinese women were taught these essentialisms and how they were (and continue to be) reproduced. This may have important implications for understanding current debates surrounding China–Australia relations.

There are also many avenues for future research if we cast our attention beyond dedicated investigations of Chinese Australian females. Firstly, in line with the work of Rose (2003), Tolia-Kelly (2004a, 2004b) and Ratnam (2018, 2020) this project has reiterated the importance of memory, storytelling and material cultures in unearthing 'hidden' histories and geographies. Thus, investigations relating to how migrant females and their descendants tell stories of their families, their homes, identities and pasts via material cultures such as photographs would help continue to correct androcentric and ethnocentric research traditions. Secondly, the findings presented in this book, and the postcolonial feminist approach used in the study, could also be extended through a comparative framework to allow 'other' women's experiences to be brought in from the margins. For example, international comparisons between Chinese Australian female experiences and Chinese female experiences in other White-settler nations such as Canada and the United States (and indeed other Chinese diaspora countries across the globe) would provide a broader yet more nuanced, understandings of Chinese female migration in the global context. If we remain local, comparisons between Chinese Australian female lives and the experiences of other migrant females in Australia and their descendants would cast light on how national discourses of 'Self' and 'Other' and its implementation in legislation and institutional practises impacted (and continue to impact) female lives in differing ways.

This could come in the form of comparing the experiences of females who were part of assisted or humanitarian migration schemes in the White Australia Policy era (e.g., Jewish, Italian and Greek migrants) with Chinese Australian females (who were stridently cast as undesirable outsiders). Thirdly, by making space for 'other' stories to be investigated, we can also extend this research to examine not only the similarities and differences, but the *intersections* of female lives, that is, the ways in which the lives of Chinese Australian females intersected with the complex histories and geographies of other marginalised groups in Australia such as Aboriginal Australians and other migrant groups (both female and male members). Perhaps these experiences were interwoven and provide further context for the experiences of Chinese Australian females (and others) in the period.

The final direction for future research that has been highlighted by this project and is important to note is the need for a more strident application of critical race theory to the analysis of female Chinese Australian experiences. Such an application may offer new understandings of racialising processes within the Australian context—in both the past and present.

References

Anderson, K. 1995, 'Culture and nature at the Adelaide Zoo: At the frontiers of 'human'geography', *Transactions of the Institute of British Geographers*, vol. 20, no. 3, pp. 275–294.

Bagnall, K. 2006, *Golden Shadows on a White Land: An Exploration of the Lives of White Women Who Partnered Chinese Men and their Children in Southern Australia, 1855–1915*, unpublished PhD thesis, University of Sydney, Sydney.

Beoku-Betts, J. A. 1995, 'We got our way of cooking things: Women, food, and preservation of cultural identity among the Gullah', *Gender & Society*, vol. 9, no. 5, pp. 535–555.

Billson, J. M. 1995, *Keepers of the Culture: The Power of Tradition in Women's Lives*, Lexington, New York.

Chan, H. 1995, 'A decade of achievement and future directions in research on the history of the Chinese in Australia', in P. Macrgegor (ed.) *Histories of the Chinese in Australiasia and the South Pacific: Proceedings of an International Public Conference Held at the Museum of Chinese Australian History, Melbourne, 8–10 October, 1993*, The Chinese Museum, Melbourne, pp. 419–423.

Chan, H. 2001, 'Becoming Australasian but remaining Chinese: The future of the down under Chinese past', in H. Chan, A. Curthoys and N. Chiang (eds.) *The Overseas Chinese in Australasia: History, Settlement and Interactions*, Centre for the Study of the Chinese Southern Diaspora, Australian National University, Canberra, pp. 1–15.

Choi, C. 1975, *Chinese Migration and Settlement in Australia*, Sydney University Press, Sydney.

Couchman, S. 2004, 'Oh I would like to see Maggie Moore again: Selected women of Melbourne's Chinatown', in S. Couchman, J. Fitzgerald and P. Macgregor (eds.) *After the Rush: Regulation, Participation and Chinese Communities in Australia 1860–1940*, Otherland Press, Melbourne, pp. 171–190.

Cushman, J. W. 1984, 'A "colonial casualty": The Chinese community in Australian historiography', *Asian Studies Association of Australia Review*, vol. 7, no. 3, pp. 100–113.

Dasgupta, S. D. 1998, 'Gender roles and cultural continuity in the Asian Indian immigrant community in the U.S', *Sex Roles*, vol. 38, no. 11–12, pp. 953–974.

Domosh, M. and Morin, K. M. 2003, 'Travels with feminist historical geography', *Gender, Place & Culture*, vol. 10, no. 3, pp. 257–264.

Ebaugh, H. R. and Chafetz, J. S. 1999, 'Agents for cultural reproduction and structural change: The ironic role of women in immigrant religious institutions', *Social Forces*, vol. 78, no. 2, pp. 585–612.

Fincher, R. 1997, 'Gender, age, and ethnicity in immigration for an Australian nation', *Environment and Planning A*, vol. 29, pp. 217–236.

Fitzgerald, J. 2007, *Big White Lie: Chinese Australians in White Australia*, University of New South Wales Press, Sydney.

Gassin, G. 2021, 'All eyes on you: Debutantes' explorations of chinese australian womanhood at the dragon festival ball', *Australian Historical Studies* (online), pp. 1–16.

Giese, D. 1997, *Astronauts, Lost Souls & Dragons: Voices of Today's Chinese Australians in Conversation with Diana Giese*, University of Queensland Press, St Lucia, Queensland.

Gleeson, B. 2001, 'Domestic space and disability in nineteenth-century Melbourne, Australia', *Journal of Historical Geography*, vol. 27, no. 2, pp. 223–240.

Gorman-Murray, A., Kamp, A. and McKinnon, S. 2018, 'Guest editorial: Historical geographies down under', *Australian Geographer*, vol. 49, pp. 1, 1–3.

Haebich, A. 2000, *Broken Circles: Fragmenting Indigenous Families 1800–2000*, Fremantle Arts Centre Press, Fremantle, W.A.

Han, E. 2011, 'Chinese Australians call for an apology', *Sydney Morning Herald (online)*, Sydney.

Honig, B. 1994, 'Difference, dilemmas, and the politics of the home', *Social Research*, vol. 61, pp. 563–597.

hooks, b. 1991, *Yearning: Race, Gender and Cultural Politics*, Turnaround, London.

Ip, M. 1990, *Home Away from Home: Life Stories of Chinese Women in New Zealand*, New Women's Press, Auckland.

Ip, M. 1995, 'From gold mountain women to astronauts' wives: Challenges to New Zealand Chinese women', in P. Macgregor (ed.) *Histories of the Chinese in Australasia and the South Pacific*, Museum of Chinese Australian History, Melbourne, pp. 274–286.

Ip, M. 2002, 'Redefining Chinese female migration: From exclusion to transnationalism', in L. Fraser and K. Pickles (eds.) *Shifting Centres: Women and Migration in New Zealand History*, Otago University Press, Dunedin, pp. 149–165.

Kamp, A., 2013, 'Chinese Australian Women in White Australia: Utilising available sources to overcome the challenge of "invisibility"', *Chinese Southern Diaspora Studies*, vol. 6, pp. 75–101.

Kay, J. 1989, 'Western women's history', *Journal of Historical Geography*, vol. 15, pp. 302–305.

Kay, J. 1990, 'The future of historical geography in the United States', *Annals of the Association of American Geographers*, vol. 80, no. 4, pp. 618–621.

Kay, J. 1991, 'Landscapes of women and men: Rethinking the regional historical geography of the United States and Canada,' *Journal of Historical Geography*, vol. 17, no. 4, pp. 435–452.

Khoo, T. and Noonan, R. 2011, 'Wartime fundraising by Chinese Australian communities', *Australian Historical Studies*, vol. 42, pp. 92–110.

Le Espiritu, Y. 2001, '"We don't sleep around like white girls do": Family, culture, and gender in Filipina American lives', *Signs*, vol. 26, no. 2, pp. 415–440.

Legg, S. 2003, 'Gendered politics and nationalised homes: Women and the anti-colonial struggle in Delhi, 1930–47', *Gender, Place & Culture*, vol. 10, no. 1, pp. 7–27.

Ling, H. 1993, 'Surviving on the gold mountain: A review of sources about Chinese American women', *The History Teacher*, vol. 26, pp. 459–470.

Ling, H. 2000, 'Family and marriage of late-nineteenth and early-twentieth century Chinese immigrant women', *Journal of American Ethnic History*, vol. 19, pp. 43–63.

Loh, M. 1986, 'Celebrating survival - an overview, 1856–1986', in Loh, M. and Ramsey, C. (eds.) *Survival and Celebration: An Insight into the Lives of Chinese Immigrant Women, European Women Married to Chinese and their Female Children in Australia from 1856–1986*, Published by the editors, Melbourne, pp. 1–10.

Lowe Kelley, D. 2011, 'Chinese Australians owed apology for discrimination against forebears', *Sydney Morning Herald* (online), Sydney.

Maddrell, A. M. C. 1998, 'Discourses of race and gender and the comparative method in geography texts 1830–1918', *Environment and Planning D: Society and Space*, vol. 16, pp. 81–103.

Mani, L. 1993, 'Gender, class, and cultural conflict: Indu Krishnan's knowing her place', In Women of South Asian Descent Collective (Eds.), *Our feet walk the sky: women of the South Asian diaspora*, Aunt Lute Press, San Francisco, CA, pp. 11–14.

Martínez, J. T. 2011, 'Patriotic Chinese women: Followers of Sun Yat-sen in Darwin, Australia', in L. Hock Guan and L. Lai To (eds) *Sun Yat-Sen, Nanyang and the 1911 Revolution*, ISEAS, Singapore. pp. 200–218.

Martínez, J. T. 2021, 'Mary Chong and Gwen Fong: University-Educated Chinese Australian Women', in K. Bagnall and J. T. Martínez (eds.) *Locating Chinese Women: Historical Mobility between China and Australia*, Hong Kong University Press, Hong Kong, pp. 204–229.

McEwan, C. 1998, 'Gender, science and physical geography in nineteenth-century Britain', *Area*, vol. 30, pp. 215–223.

McKewon, E. 2003, 'The historical geography of prostitution in Perth, Western Australia', *Australian Geographer*, vol. 34, no. 3, pp. 297–310.

Mohammad, R. 2007, 'British Pakistani Muslim women: Marking the body, marking the nation', in L. Nelson and J. Seager (eds.) *A Companion to Feminist Geography*, Blackwell Publishing Ltd, Oxford, pp. 379–397.

Morin, K. M. and Berg, L. D. 1999, 'Emplacing current trends in feminist historical geography', *Gender, Place & Culture*, vol. 6, no. 4, pp. 311–330.

Phillips, R. 1997, *Mapping Men and Empire: A Geography of Adventure*, Routledge, London.

Ramsay, G. 2003, 'Cherbourg's Chinatown: Creating an identity of place on an Australian Aboriginal settlement', *Journal of Historical Geography*, vol. 29, no. 1, pp. 109–122.

Ratnam, C. 2018, 'Creating home: Intersections of memory and identity', *Geography Compass*, vol. 12, no. 4, pp. 1–11.

Ratnam, C. 2020, '(Re)creating home: The lived and gendered experiences of tamil women in Sydney, Australia', in N. Kandasamy, N. Perera and C. Ratnam (eds.) *A Sense of Viidu: The (Re)creation of Home by the Sri Lankan Tamil Diaspora in Australia*, Springer, Singapore, pp. 119–139.

Rose, G. 2003, 'Family photographs and domestic spacings: A case study', *Transactions of the Institute of British Geographers*, New Series, vol. 28, no. 1, pp. 5–18.

Rose, G. and Ogborn, M. 1988, 'Feminism and historical geography', *Journal of Historical Geography*, vol. 14, pp. 405.

Schuurman, N. 1998, 'Contesting patriarchies: Nlha7pamux and Stl'atl'imx women and colonialism in nineteenth-century British Columbia', *Gender, Place & Culture*, vol. 5, no. 2, pp. 141–158.

Song, M. 1995, 'Between "the front" and "the back": Chinese women's work in family businesses', *Women's Studies International Forum*, vol. 18, no. 3, pp. 285–298.

Tan, C. 2003, 'Living with 'difference': Growing up 'Chinese' in white Australia', *Journal of Australian Studies*, vol. 77, pp. 101–108, 195–197.

Teather, E. K. 1990, 'Early postwar Sydney: A comparison of its portrayal in fiction and in official documents', *Australian Geographical Studies*, vol. 28, no. 2, pp. 204–223.

Teather, E. K. 1992, 'The first rural women's network in New South Wales: Seventy years of the Country Women's Association', *Australian Geographer*, vol. 23, no. 2, pp. 164–176.

Teng, J. E. 1996 'The construction of the "traditional Chinese woman" in the Western Academy: A critical review', *Signs*, vol. 22, no. 1, pp. 115–151.

Tolia-Kelly, D. P. 2004a, 'Locating processes of identification: Studying the precipitates of re-memory through artefacts in the British Asian home', *Transactions of the Institute of British Geographers*, vol. 29, no. 3, pp. 314–329.

Tolia-Kelly, D. P. 2004b, 'Materializing post-colonial geographies: Examining the textural landscapes of migration in the South Asian home', *Geoforum*, vol. 35, no. 6, pp. 675–688.

Watson, R. S. and Ebrey, P. B. (eds.) 1991, 'Marriage and inequality in Chinese society', *Studies on China*, vol. 12, University of California Press, Los Angeles, CA.

Yong, C. 1977, *The New Gold Mountain: The Chinese in Australia 1901–1921*, Raphael Arts, Adelaide.

Young, I. M. 1997, *Intersecting Voices: Dilemmas of Gender, Political Philosophy, and Policy*, Princeton University Press, Princeton, NJ.

Yung, J. 1998, 'Giving voice to Chinese American women', *Frontiers: A Journal of Women Studies*, vol. 19, pp. 130–156.

Appendix A

Brief biographies of interview participants

Table A.1 Interview Participant Information[1]

Name[2]	Year of Birth	Place of Birth	Generation[3]	Year of Ancestral Migration/Self-migration[4]	Highest Level of Education/Qualification	Occupational Background	Marital Status	# children
Mabel	1939	Warialda, NSW, Australia	Second (Australian-born father; Chinese-born mother)	c.1880	Postgraduate degree	Academic	Married	1
Sandy	1943	Hong Kong	First	1962	Diploma	High School Teacher; Information Technology	Married	2
Ina	1943	Sydney, NSW, Australia	Second (Australian-born father; Chinese-born mother)	c.1890	Bachelor's degree	High School Teacher	Married	2
Lily	1939	Hong Kong	First	1971	Postgraduate degree	Academic; Librarian	Married	4
Susan	1938	Sydney, NSW, Australia	Second (Australian-born father; Chinese-born mother)	c.1900	High School (Intermediate level)	School secretary	Widowed	4
Mei-Lin	1942	Texas, QLD, Australia	Third	c.1890	Diploma	Early childhood teacher	Married	–
Daphne	1939	Wellington, New Zealand	First	1964	Bachelor's degree	High school/tertiary teacher; Travel and tourism	Married	1
Stella	1931	Thursday Island, NSW, Australia	Second (Australian-born mother; Chinese-born father)	c.1900	High School	ABC administration support staff, news department and news archives; news typist; family general store manager	Widowed	5
Sally	1942	Tumut, NSW, Australia	Second (Chinese-born father; Chinese-born mother)	1926	Tertiary Certificate	Nurse	Married	–
Nancy	1941	Sydney, NSW, Australia	Second (Australian-born father; Chinese-born mother)	c.1900	Tertiary Certificate	Midwife; child and family healthcare nurse	Married	4
Mary	1952	Hong Kong	First	1967	Diploma	Executive Secretary; Public Relations Officer; Banqueting Officer; Shiatsu therapist	Divorced	1

(Continued)

Table A.1 (Continued)

Name[2]	Year of Birth	Place of Birth	Generation[3]	Year of Ancestral Migration/Self-migration[4]	Highest Level of Education/Qualification	Occupational Background	Marital Status	# children
Doreen	1941	Sydney, NSW, Australia	Second (Chinese-born father; Chinese-born mother)	c.1920	Bachelor's degree	Book Keeping Machine Operator; Office Manager/Secretary; Business partner; Management consultant	Divorced	2
Kaylin	1923	Sydney, NSW, Australia	Second (Australian-born mother; Chinese-born father)	1868	Tertiary Certificate	Company Secretary; Accountant	Single	–
Marina	1934	Inverell, NSW, Australia	Third	c.1892	High School (Intermediate level)	Ladies retail	Widowed	2
Edith	1949	Glenn Innes, NSW, Australia	Third	1885	Tertiary Certificate	Secretary; Business owner	Married	3
Eileen	1946	Sydney, NSW, Australia	Second (Australian-born father; Chinese-born mother)	c.1896	High School (Intermediate level)	Clerk/Secretary	Married	5
Patricia	1941	Dongguan Guangdong, China	First	1947	Bachelor's degree	Microbiologist	Married	3
Margaret	1948	Sydney, NSW, Australia	Second (Chinese-born father; Chinese-born mother)	c.1946	Master's degree	Tax assessor/auditor	Single	–
Helen	1943	Hong Kong	First	1961	Bachelor's degree	Social worker; TAFE teacher; Solicitor; Legislator	Married	2

1 This information was obtained at the time of interviews and may have changed.

2 Participants' names are used with consent. Where consent was not granted, pseudonyms have been applied.

3 In accordance with definitions of 'generation' used by the Australian Bureau of Statistics, 'first generation Australians' are people living in Australia who were born overseas; 'second generation Australians' are Australian-born people living in Australia, with at least one overseas-born parent; 'third-plus generation Australians' are Australian-born people whose parents were both born in Australia—one or more of their grandparents may have been born overseas or they may have several generations of ancestors born in Australia (ABS 2013).

4 For participants that are migrant/first generation Australians, the year of their arrival to Australia is presented. For those participants that are second-plus generation Australians, the year that the first of their forbears arrived in Australia is presented.

Mabel Lee is second generation Australian. Her paternal grandfather, a merchant, came to Australia from Guangdong Province, China in the late 1880s. Her paternal grandmother followed him soon after. Mabel's father was the youngest of four children all born in Sydney. Mabel's mother was born in Guangdong province in 1903. Mabel's parents were married in their ancestral village in 1922. After bearing three children in China, Mabel's mother migrated to Australia in 1939 to be reunited with her husband. Mabel, who is one of seven children, was born soon after in 1939 in Warialda, a small town in rural New South Wales, Australia. Mabel moved from Warialda to the western suburbs of Sydney with her family at the age of five as her father bought a farm and later opened a fruit shop. After completing her high school education, Mabel went on to complete her PhD in Chinese studies at Sydney University in the mid-1960s. At university, she was the President of the Chinese Student's Association. Mabel was first married in 1973 to a Chinese-born Australian and has one daughter. She is now married to a British Australian and continues her academic work in Chinese intellectual history and literature at Sydney University.

Sandy is first generation Australian. Born in Hong Kong in 1944, Sandy is one of six children (four girls and two boys). Sandy joined her older sister in Melbourne, Australia, in 1962 to complete her high-school education. She went on to complete a Diploma in Maths and Secondary Teaching and has worked as a high school teacher and in IT. Sandy married her Australian husband in 1970 and they have one daughter and one son. Sandy is now retired and spends her time volunteering for the Older Women's Network and babysitting her granddaughter.

Ina is second generation Australian. Born in 1943 in Sydney, Australia, Ina was one of four children (having two sisters and one brother). Ina's father was born in the mining town of Cobar in regional New South Wales and spent some of his teenage years in China to obtain a Chinese education. As an adult, he returned to China to marry Ina's mother. After a ten-year separation, Ina's mother migrated to Australia to reunite with her husband, leaving behind her first-born child. Ina's father ran a fruit and vegetable shop on Sydney's Parramatta Road and her mother helped in the business, cleaning and cooking for staff. Ina completed her high school education and went on to finish a Bachelor of Arts (in English and History) and a Diploma of Education. She has been a high school teacher in Sydney, Amsterdam, London and Hong Kong. She is married to a Chinese

Indonesian and they have two sons. Ina is now retired from teaching and tutors English to high school students.

Lily Xiao Hong Lee is first generation Australian and was born in Hong Kong in 1939. Due to the impacts of the Chinese Cultural Revolution and a desire for a better life for their children, Lily and her husband migrated to Sydney, Australia, from Hong Kong with their two children in 1971. Lily and her husband had an additional two children after their arrival in Australia. Lily obtained a Bachelor of Arts in Singapore, Masters of Library Science in the United States and worked as a librarian in Hong Kong and Australia. While working as the Oriental Librarian at Sydney University, Lily completed her PhD in Chinese Studies. Lily is an Honorary Associate at Sydney University and has completed a Biographical Dictionary of Chinese Women.

Susan is second generation Australian. Her Australian-born father was born in Melbourne but returned to China where he married Susan's Chinese-born mother. After their arrival in Australia in 1934, Susan's parents established a fruit shop in Parramatta in Sydney's west and started their family. Born in 1938, Susan is one of four children (two girls and three boys). Susan completed her high school education at the intermediate level (at approximately 15 years of age) and went on to become a legal secretary until she married her late Chinese-Malaysian husband at the age of 22 in 1960. After her husband passed away, Susan became a school secretary at a prestigious private Sydney girls' school to support her four children. After working as a school secretary for 30 years, Susan was able to retire close to her 68th birthday. She now spends her time exercising, volunteering in a crèche and driving the elderly to appointments. She is a member of the Mater Hospital fundraising committee and the University of the Third Age. Susan also enjoys going to the ballet, having computer lessons and playing Bridge and Mahjong.

Mei Lin Yum is third generation Australian. Her maternal great-grandmother, Sarah, arrived in Australia from China in the nineteenth century. Mei-Lin's maternal grandmother (whose father was White Australian)

and mother were subsequently born in Australia. Mei-Lin's father was also Australian-born. His parents had migrated to Australia from Guangdong Province, China in the 1890s and had 12 children after settling in The Rocks, and then Marrickville in Sydney. Mei-Lin's parents were married in 1934 and Mei-Lin was born in 1942 in Texas, Queensland, the first of two children (she has one brother). During her early childhood, her father managed the Texas branch of Harry Fay's Hong Yuen store. In 1954, the family moved to Ballina, a small town in northern New South Wales, where Mei-Lin's parents obtained ownership of their own store and the children had greater educational opportunities. After completing high school, Mei-Lin attended Teacher's College in Armidale, New South Wales, and became an Early Childhood Teacher in Sydney where she has remained ever since. Mei-Lin and her late partner John never married. She has been retired since 2002 and is a member of the Chinese Women's Association.

Daphne Lowe Kelley is first generation Australian. Born in New Zealand to migrant Chinese parents, Daphne is one of seven children. Daphne migrated to Australia in 1964 to live with her first husband, a Chinese Australian with whom she has one son. With a university degree in teaching, Daphne taught in high schools for five years before teaching at TAFE for nearly 15 years. Later in life, with a Graduate Diploma in Tourism Management, Daphne changed occupations and entered the travel and tourism industry. Daphne spends her retirement involved in a host of Chinese and non-Chinese associations including the Chinese Women's Association, the Australian Branch of the International Institute for Peace through Tourism, and the Chinese Heritage Association of Australia of which she was president for ten years. Daphne is an ardent advocate of an official government apology for the discrimination and mistreatment of Chinese in Australia's past and the preservation of important heritage sites such as surviving market gardens in Sydney's south. Through theatre and arts events, Daphne also encourages a broader understanding and appreciation of Australia's Chinese cultural heritage.

Stella Sun is second generation Australian. Her maternal grandmother was sold as child in China and via an arranged marriage, came to Australia to be the concubine of a general merchant and pearl dealer on Thursday Island, off the coast of far north Queensland. Stella's mother was born in 1906 on Thursday Island, the first of nine children. Stella's mother was sent to China to be married but soon became estranged from her husband. Stella was born in 1931 and was raised by her grandparents at their general store on Thursday Island. At the age of 16, Stella managed one of her grandfather's small stores. However, eager to leave Thursday Island, Stella moved to Sydney and with the help of an editor of a local paper and the Young Women's Christian Association. After working briefly as a news typist at the Sydney Morning Herald, Stella spent the rest of her working life at the ABC in administration, in the news department and news archives. Stella married her late Chinese Australian husband when she was 21 years of age. She has three daughters and two sons.

Sally Pang is second generation Australian with her ancestral village situated in Guangdong Province, China. Her father arrived in Sydney from Hong Kong in 1926 to manage a trading store Sally's grandfather had previously established. Sally's mother and one of her children arrived in 1933, followed by the remaining three Hong-Kong-born children. An additional three children were born in Australia, including Sally who was born in Tumut in regional New South Wales in 1942. Sally grew up in Sydney's inner-west, spending a lot of her time at her father's Modern China Café in Chinatown. After finishing school, Sally began a university degree before deciding nursing was her calling. Between 1971 and 1997, Sally worked in the health care system—as a nurse and as a continence advisor. Sally established the first continence advisory group at Royal Prince Alfred Hospital in the late 1980s as well as the first Toastmasters group at the hospital. Sally married her husband in 1988. In her retirement, Sally volunteers for the Continence Foundation of Australia and Toastmasters. She is also a member of the Australian Chinese Community Association and is involved with the Australian Chinese Charity Foundation and the Chinese Heritage Association of Australia.

Nancy Buggy is second generation Australian. Her paternal grandfather arrived in Australia at the turn of the twentieth century and her father was born in Melbourne in 1908/1909. Nancy's mother was a migrant, who arrived in Australia from China in the 1930s. The marriage of Nancy's parents was arranged and took place in China. The couple settled in Sydney (after a brief stay in Boggabri, a small town in north-western New South Wales) and established a fruit and vegetable shop in the western suburbs where their children spent their childhood. After completing her high school education, Nancy became a nurse specialising in midwifery. After working for four years, Nancy packed her bags and travelled to Canada. It was on the ship to Canada that she met her late husband, an Irish Australian. The couple married in London and returned from abroad after three years. Nancy and her husband have four children and Nancy is retired from her nursing career. She spends her retirement learning French, doing Tai-Chi, looking after her grandchildren and volunteering at the local library and in English language classes.

Mary Tang is first generation Australian. Mary was born in Hong Kong in 1942, the fourth child of her father's second wife. As a female, her birth was considered to be a bad omen for the family and she was subsequently given away as a baby. After several years, her foster parents were no longer able to care for her and Mary was returned to her birth family at the age of five. In 1967, at the age of 14, Mary was sent to Australia with her sister to complete her education. After completing her high school education and a Diploma in Office Administration, Mary returned to Hong Kong for two years and studied hotel management. In 1974, Mary decided to return to Australia and married in 1976. She has one son and is now divorced. Mary has worked as an Executive Secretary, Public Relations Officer in aircraft catering, a Banqueting Officer in hotels and a Shiatsu Therapist. In 2002, Mary was introduced to poetry writing and now produces work in English and Chinese. Mary also teaches Chinese Calligraphy as a volunteer tutor for the University of the Third Age.

Doreen Cheong is second generation Australian. Doreen's paternal great-grandfather came to Australia from Guangdong Province in the 1890s. However, after improving his economic circumstances he returned to China and his family. Doreen's paternal grandfather came to Australia in the early 1900s, leaving his wife and only son (Doreen's father) behind in China. Doreen's father arrived in the 1920s and like his father, left his wife and children behind. He returned to China in the mid-1930s and before returning once again to Sydney, Australia, took on a second wife (Doreen's mother). In the late 1930s, Doreen's mother was granted entry into Australia to reunite with her husband. Doreen was born in 1941. Doreen spent much of her childhood growing up at her father's market garden in Sydney's southwest.

Doreen married a Chinese migrant in 1961 but the couple separated after 17 years. They had two children. As a young mother, Doreen completed her high school education and in 1979 completed a Bachelor of Arts in Psychology and Economics. She has worked as a book-keeping machine operator, office manager/secretary, as a partner in her ex-husband's jewellery business and a management consultant. Doreen has spent her retirement supervising second-year Master of Organisational Psychology students at the University of New South Wales, learning Cantonese, attending art appreciation/history lectures, the theatre and loves to travel.

Kaylin Simpson Lee was second generation Australian. Her maternal grandfather came to Australia in 1868 and, with the help of his wife who later joined him in Australia, established the Presbyterian Church in Sydney. Kaylin's mother was one of six Australian-born children. Kaylin's father migrated to Australia in 1900 and first lived in Perth, Western Australia. He later came to Sydney and joined the Presbyterian Church—which is where he met Kaylin's mother. Kaylin's father, with some of his compatriots, established a furniture factory and then a retail furniture business in the Central Business District of Sydney of which he was the principal. Kaylin was born in 1923 and grew up in the suburbs of Sydney, quite isolated from the rest of Sydney's Chinese community. After completing her high school education, Kaylin successfully passed the examinations of the Chartered Institute of Secretaries. Kaylin worked as the assistant company secretary of her father's furniture business and as the accountant and company secretary for a manufacturing firm that was involved in designing and printing fabrics. Kaylin was actively involved in various professional women's organisations including the Australian National Council of Women of NSW (of which she served on the executive), the Pan Pacific and South East Asia Women's Association, and The Women's Club in Sydney (of which she served on the Board of Directors, was the President for one term before becoming an Honorary Life Member). She was also a member of the Chinese Women's Association and Country Women's Association. Kaylin retired in 1983 and continued to be involved in these organisations. Kaylin passed away in February 2021.

Marina Mar is third generation Australian. She was born in 1934 in Inverell, a small town in northern New South Wales. Her father, Harry Fay, was also Australian-born, however spent much of his childhood and adolescent years in China. On his return to Australia, Harry Fay worked in various Chinese-owned stores in northern New South Wales, before eventually gaining ownership of the famous Hong Yuen department store in Inverell. Marina's mother was born in 1897 in Sydney and was of mixed Chinese descent—her father was Chinese and her mother was English. Marina's maternal grandmother suffered from mental illness and Marina's mother was subsequently adopted by Mr Wong Chee of Glen Innes. It was through Mr Wong Chee that Marina's mother and father met and were married in 1916. Marina spent her childhood with her seven siblings in Inverell and began working at Hong Yuen at the age of 15. She eventually became the manager of the ladies department and after her marriage in 1962 moved to Sydney where her Chinese Australian husband worked as an architect. Marina has two sons and is a member of the Chinese Women's Association. Marina currently attends Mandarin and computer classes, is learning to play the guitar and is a volunteer at a nursing home.

Edith is third generation Australian. Both her maternal and paternal grandparents arrived in Australia from China in the 1880s in search of a better life. Her paternal grandparents settled in Glen Innes, in regional New South Wales, while her maternal grandparents settled on the New South Wales south coast. Her parents met in Port Kembla and settled in Glen Innes where Edith's father was a partner in the family business. Edith is one of four children and at the age of eight, her family relocated to Sydney where her father set up a grocery store. After leaving high school, Edith completed a secretarial course and worked as a secretary in an accounting firm. In 1978, Edith was married to a Chinese-born Australian and together they ran an engineering company. They have three children and Edith is now retired.

Eileen Yip is second generation Australian. Her maternal great grandfather was a migrant from China who arrived in Australia in the nineteenth century. Her maternal great grandmother was a Scottish migrant. Eileen's Australian-born grandmother was born in 1896 in Queensland. She married Eileen's grandfather, a Chinese migrant from Canton, in 1912. In 1914, Eileen's father was born in the Atherton Tablelands in northern Queensland. Eileen's mother was a Chinese migrant who arrived in Australia in the early twentieth century as the wife of her first husband, a Chinese herbalist. After his passing, Eileen's parents were married in 1945. Eileen was born in Sydney in 1946. Eileen spent most of her childhood in Cairns where her parents established a grocery store and later a fish and chips shop. After leaving high school at the intermediate level, Eileen worked at the Ford Motor Company in Cairns until she was married (at age 20). After her marriage, she moved to Sydney where her husband owned a produce company. Eileen has five children (five sons and one daughter) and actively participated in their schooling, being the President of the parent's committee for seven years. Eileen now enjoys line-dancing and caring for her grandchildren.

Patricia Yip is first generation Australian. Born in Guangdong Province, China, in 1941 Patricia migrated to Australia with her family at the age of six in 1947. Patricia's father had arrived years earlier, being sponsored by his brother-in-law (his sister's husband) to help in their restaurant business. The family settled in Surry Hill, Sydney. After completing high school, Patricia undertook a Bachelor of Science at Sydney University and with her qualifications worked in medical research. Patricia later helped her husband, a Chinese Australian, in his Pharmacy before becoming a full-time mother to three children. Patricia now spends her time learning Mandarin, line-dancing, and looking after her grandchildren.

Margaret is second generation Australian. Margaret's Chinese-born parents left China for New Guinea at the onset of the second Sino-Japanese War, and later left New Guinea for Australia at the outbreak of World War II. Settling in the western suburbs of Sydney, Margaret's parents established a fruit shop. Margaret was born in 1948 in Sydney and is one of five children. Margaret spent much of her childhood and adolescence helping her parents in their family business. After she completed high school, she attended the Metropolitan Business College for secretarial training and worked as a tax assessor/auditor until her retirement. Margaret also has a Bachelor of Social Science and a Masters of Tax. She never married and enjoys her retirement attending classes held by the Workers Education Association, and is a member of various museums and art galleries in Sydney.

Helen Sham-Ho is first generation Australian. Born in Hong Kong in 1943, the fifth of six children, Helen came to Australia in 1961 to complete her high school education in Sydney. She later completed a Bachelor of Arts and Diploma of Social Work at Sydney University where she was the representative of the Overseas Student Association. As a social worker, Helen worked in various hospitals in Sydney. She married her Chinese Australian husband in 1968 and had two children but was divorced in 1978. Helen completed a Bachelor of Legal Studies at Macquarie University in 1985 while looking after her two children and teaching Social Welfare at TAFE. She worked as a solicitor and became active in the Chinese community in Sydney. Helen re-married in 1987 and one year later was a member of the New South Wales Legislative Council for the Liberal Party. Helen resigned from the Liberal Party in 1998. Helen was the first Chinese-born parliamentarian in Australia. In her retirement, Helen continues to make important contributions to Chinese and non-Chinese community organisations.

Appendix B
List of census publications

1911. Volume II. Part VIII. Non-European Races

- No. 2. Persons of non-European race enumerated in the Commonwealth of Australia. Summary by Races.
- No. 7. Commonwealth of Australia–Females of non-European race enumerated. Classified according to Race and Age.
- No. 35. Commonwealth of Australia–Females of non-European race enumerated. Classified according to Race, Nationality, Length of Residence, Education and Conjugal Condition.
- No. 63. Commonwealth of Australia–Females of non-European race enumerated. Classified according to Birthplace and Race.
- No.90. Commonwealth of Australia–Females of non-European race enumerated. Classified according to Occupation and Race.
- No. 113. Females of non-European race enumerated in the metropolitan areas of the several states of the Commonwealth of Australia.

1921. Volume I. Part V. Race

- No. 1. Numbers, according to Race, of Males and Females born in Australia recorded in Urban and Rural Divisions throughout Australia at the Census of the 4th April 1921.
- No. 2. Numbers, according to Race, of Males and Females born outside Australia recorded in Urban and Rural Division throughout Australia at the Census of the 4th April 1921.
- No. 3. Numbers, according to Race, of Males and Females (including those whose birthplace was unstated) recorded in Urban and Rural Division throughout Australia at the Census of the 4th April 1921.
- No. 4. Numbers, according to Race, of Males and Females born in Australia recorded in Urban and Rural Division of New South Wales at the Census of the 4th April 1921.
- No. 5. Numbers, according to Race, of Males and Females born outside Australia recorded in Urban and Rural Division of New South Wales at the Census of the 4th April 1921.

- No. 6. Numbers, according to Race, of Males and Females (including those whose birthplace was unstated) recorded in Urban and Rural Division of New South Wales at the Census of the 4th April 1921.
- No. 7. Numbers, according to Race, of Males and Females born in Australia recorded in Urban and Rural Division of Victoria at the Census of the 4th April 1921.
- No. 8. Numbers, according to Race, of Males and Females born outside Australia recorded in Urban and Rural Division of Victoria at the Census of the 4th April 1921.
- No. 9. Numbers, according to Race, of Males and Females (including those whose birthplace was unstated) recorded in Urban and Rural Division of Victoria at the Census of the 4th April 1921.
- No. 10. Numbers, according to Race, of Males and Females born in Australia recorded in Urban and Rural Division of Queensland at the Census of the 4th April 1921.
- No. 11. Numbers, according to Race, of Males and Females born outside Australia recorded in Urban and Rural Division of Queensland at the Census of the 4th April 1921.
- No. 12. Numbers, according to Race, of Males and Females (including those whose birthplace was unstated) recorded in Urban and Rural Division of Queensland at the Census of the 4th April 1921.
- No. 13. Numbers, according to Race, of Males and Females born in Australia recorded in Urban and Rural Division of South Australia at the Census of the 4th April 1921.
- No. 14. Numbers, according to Race, of Males and Females born outside Australia recorded in Urban and Rural Division of South Australia at the Census of the 4th April 1921.
- No. 15. Numbers, according to Race, of Males and Females s (including those whose birthplace was unstated) recorded in Urban and Rural Division of South Australia at the Census of the 4th April 1921.
- No. 16. Numbers, according to Race, of Males and Females born in Australia recorded in Urban and Rural Division of Western Australia at the Census of the 4th April 1921.
- No. 17. Numbers, according to Race, of Males and Females born outside Australia recorded in Urban and Rural Division of Western Australia at the Census of the 4th April 1921.
- No. 18. Numbers, according to Race, of Males and Females (including those whose birthplace was unstated) recorded in Urban and Rural Division of Western Australia at the Census of the 4th April 1921.
- No. 19. Numbers, according to Race, of Males and Females born in Australia recorded in Urban and Rural Division of Tasmania at the Census of the 4th April 1921.
- No. 20. Numbers, according to Race, of Males and Females born outside Australia recorded in Urban and Rural Division of Tasmania at the Census of the 4th April 1921.

- No. 21. Numbers, according to Race, of Males and Females (including those whose birthplace was unstated) recorded in Urban and Rural Division of Tasmania at the Census of the 4th April 1921.
- No. 22. Numbers, according to Race, of Males and Females born in Australia recorded in Urban and Rural Division of the Northern Territory and in the Federal Capital Territory at the Census of the 4th April 1921.
- No. 23. Numbers, according to Race, of Males and Females born outside Australia recorded in Urban and Rural Division of Northern Territory and in the Federal Capital Territory at the Census of the 4th April 1921.
- No. 24. Numbers, according to Race, of Males and Females (including those whose birthplace was unstated) recorded in Urban and Rural Division of Northern Territory and in the Federal Capital Territory at the Census of the 4th April 1921.
- No. 26. Australia. Females classified according to Race in conjunction with Age and Grade of Occupation.

1933. Volume I. Part XII. Race

- No.1. Males in each State and Territory classified according to Race as recorded at the Census of the 30th June 1933.
- No. 2. Females in each State and Territory classified according to Race as recorded at the Census of the 30th June 1933.
- No. 5. Males and Females Born in Australia recorded in Urban and Rural Divisions at the Census of the 30th June 1933.
- No. 6. Males and Females Born outside Australia recorded in Urban and Rural Divisions at the Census of the 30th June 1933.
- No. 7. Males and Females recorded in Urban and Rural Divisions at the Census of the 30th June 1933.
- No. 8. New South Wales. Males and Females recorded in Urban and Rural Divisions at the Census of the 30th June 1933.
- No. 9. Victoria. Males and Females recorded in Urban and Rural Divisions at the Census of the 30th June 1933.
- No. 10. Queensland. Males and Females recorded in Urban and Rural Divisions at the Census of the 30th June 1933.
- No. 11. South Australia. Males and Females recorded in Urban and Rural Divisions at the Census of the 30th June 1933.
- No. 12. Western Australia. Males and Females recorded in Urban and Rural Divisions at the Census of the 30th June 1933.
- No. 13. Tasmania. Males and Females recorded in Urban and Rural Divisions at the Census of the 30th June 1933.
- No. 14. Territories. Males and Females recorded in Urban and Rural Divisions of the Federal Capital Territory and Northern Territory at the Census of the 30th June 1933.
- No. 16. Australia. Females classified according to Race in conjunction with Age and Conjugal Condition at the Census of the 30th June 1933.

- No. 38. Australia. Females classified according to Race in conjunction with Schooling; and Orphanhood of Females under Sixteen Years of Age at the Census of the 30th June 1933.
- No. 41. Australia. Females classified according to Race in conjunction with Grade of Occupation at the Census of the 30th June 1933.

1933. Part XIII. Period of Residence

- No. 27. Australia. Period of Residence in Australia of Females Born outside Australia classified according to Race, as recorded at the Census of the 30th June 1933.

1933. Volume II. Part XXVIII. Income

- No. 13. Australia. Female Breadwinners classified according to Income in conjunction with Race at the Census of the 30th June 1933.

1947. Volume I. Part XV. Race

- No. 7. Australia. Males and Females recorded in Urban and Rural Divisions classified according to Race: Census, 30th June 1947.
- No. 8. New South Wales. Males and Females recorded in Urban and Rural Divisions classified according to Race: Census, 30th June 1947.
- No. 9. Victoria. Males and Females recorded in Urban and Rural Divisions classified according to Race: Census, 30th June 1947.
- No. 10. Queensland. Males and Females recorded in Urban and Rural Divisions classified according to Race: Census, 30th June 1947.
- No. 11. South Australia. Males and Females recorded in Urban and Rural Divisions classified according to Race: Census, 30th June 1947.
- No. 12. Western Australia. Males and Females recorded in Urban and Rural Divisions classified according to Race: Census, 30th June 1947.
- No. 13. Tasmania. Males and Females recorded in Urban and Rural Divisions classified according to Race: Census, 30th June 1947.
- No. 14. Northern Territory and Australian Capital Territory. Males and Females recorded in Urban and Rural Divisions classified according to Race: Census, 30th June 1947.
- No. 16. Australia. Females classified according to Race in conjunction with Age: Census, 30th June 1947.

1947. Part XII. Birthplace

- No. 33. Australia. Females classified according to Birthplace in conjunction with Race: Census, 30th June 1947.

1947. Part XIII. Period of Residence in Australia of Persons Born Outside Australia

- No. 27. Australia. Period of Residence in Australia of Females Born outside Australia classified according to Race: Census, 30th June 1947.

1954. Volume VIII. Australia. Supplement to Part I. Cross-classification of the characteristics of the population: Race

- No. 1. Population according to Race: States and Territories of Australia, Census, 30th June 1954.
- No. 3. Males and Females born in and born outside Australia according to Race: Metropolitan Urban, Other Urban and Rural, etc. Divisions of Australia, Census, 30th June 1954.
- No. 5. Females classified according to Race in conjunction with Age (five year groups) and Conjugal Condition: Australia, Census, 30th June 1954.
- No.7. Females classified according to Birthplace in conjunction with Race: Australia, Census, 30th June 1954.
- No. 9. Females born outside Australia classified according to Race in conjunction with Period of Residence in Australia, Census, 30th June 1954.

1961. Volume I. New South Wales. Part II. Cross-classification of the characteristics of the population

- No. 34. Males and Females born in and born outside Australia according to Race: Metropolitan Urban, Other Urban and Rural, etc. Divisions of New South Wales, Census, 30th June 1961.

1961. Volume II. Victoria. Part II. Cross-classification of the characteristics of the population

- No. 34. Males and Females born in and born outside Australia according to Race: Metropolitan Urban, Other Urban and Rural, etc. Divisions of Victoria, Census, 30th June 1961.

1961. Volume III. Queensland. Part II. Cross-classification of the characteristics of the population

- No. 34. Males and Females born in and born outside Australia according to Race: Metropolitan Urban, Other Urban and Rural, etc. Divisions of Queensland, Census, 30th June 1961.

1961. Volume IV. South Australia. Part II. Cross-classification of the characteristics of the population

- No. 34. Males and Females born in and born outside Australia according to Race: Metropolitan Urban, Other Urban and Rural, etc. Divisions of South Australia, Census, 30th June 1961.

1961. Volume V. Western Australia. Part II. Cross-classification of the characteristics of the population

- No. 34. Males and Females born in and born outside Australia according to Race: Metropolitan Urban, Other Urban and Rural, etc. Divisions of Western Australia, Census, 30th June 1961.

1961. Volume VI. Tasmania. Part II. Cross-classification of the characteristics of the population

- No. 34. Males and Females born in and born outside Australia according to Race: Metropolitan Urban, Other Urban and Rural, etc. Divisions of Tasmania, Census, 30th June 1961.

1961. Volume VII. Territories. Part I. Northern Territory: Population

- No. 34. Males and Females born in and born outside Australia according to Race: Other Urban and Rural, etc. Divisions of the Northern Territory, Census, 30th June 1961.

1961. Volume VII. Territories. Part III. Australian Capital Territory: Population

- No. 34. Males and Females born in and born outside Australia according to Race: Other Urban and Rural, etc. Divisions of Australian Capital Territory, Census, 30th June 1961.

1961. Volume VIII. Australia. Part I. Cross-classification of the characteristics of the population

- No. 34. Males and Females born in and born outside Australia according to Race: Metropolitan Urban, Other Urban and Rural, etc. Divisions of Australia, Census, 30th June 1961.
- No. 36. Females according to Race and Age (five year groups): Australia, Census, 30th June 1961.
- No. 37. Males and Females according to Race and Conjugal Condition: Australia, Census, 30th June 1961.
- No. 39. Females according to Race and Birthplace: Australia, Census, 30th June 1961.
- No. 41. Females born outside Australia according to Race and Period of Residence in Australia: Australia, Census, 30th June 1961.

1961. Volume I. Population: Single Characteristics. Part II. Race

- No. 2. Population by Race, Urban and Rural, States and Territories: Australia, Census, 30th June 1966.

Index

Pages in *italics* refer figures, **bold** refer tables and pages followed by n refer notes.

age groups of Chinese Australian females, 1911–1966, **62**
American imperialism 32
Amoy 12, 21n7
Anderson, K. 31, 43n3, 176
Ang, I. 150, 153, 159–161
Anglo-Indian migrant women 28
anti-Chinese riots: Bendigo Field, Victoria 13; Lambing Flats, New South Wales 13
anti-Chinese sentiments 3
assimilation 15, 19–20, 115–117, 119–124, 126, 129, 135, 140, 159–162, 174–175
Atie, R. 167n1
Australian: adopting traditions 139–140; 'Australian' traditions, Christmas and Easter 139–140; culture 118; Empire Day 133, *134*; food 127; identity/ Australianness 19–20, 118, 126, 144–145, 147, 149, 151–152, 159–161, 176; immigration law 7, 15, 153; jobs 88; society 98, 118; way of life 15, 115
Australian-born: children 57; Chinese 5, 73; female children 61, **62**, 62–63; and foreign-born Chinese females, 1911–1966 **53**; generations of women 64–66, 65–68; international mobility of females 75–77; interview participants 36, 57, 64
Australian Broadcasting Corporation 155

Bagnall, K. 5, 21n4, 61, 110, 119, 126, 158
Barnett, C. 41
Bauman, Z. 160
Berg, L. D. 29, 42
Bhabha, H. 32
Big White Lie 5
Blunt, A. 27–28, 36

British women 28
Burton, A. M. 43n2

Cairns 2, 81–82, 85n7, 102, *103*, 128, 191–192
Cant, G. 41
Carbado, D. W. 32
Catholic School 156
census 5, 9, 16, 26, 54, 56, 172; colonial 44n5; data 18, 51, 61, 63, 69, 78, 80, 83, 99, 107, 110, 118–119; historical data, re-examining 26, 39–42; national 39, 44n6, 177; publications list 194–199; race definitions 44n6; state 44n5, **80**; years 52, **52**, **53**, 56, 60, 77, 79, **80**, 106, 170–171
Certificate Exempting from Dictation Test (CEDT) 15, 177
Certificate of Domicile 15
childhood memories: experiences of belonging/inclusion 19, 157, 159, 161–166; school-yard bullying 145–149
China–migration policy 84n1
Chinatown 102–103, 116–117, 127, 135, 141n2, 188
Chinese: ancestry 16, 39; arrivals and departures to/from Australia 69, **70–71**; Church 157; costume, *cheongsams* 133, *134*; culture 19, 117, 130, 132–133, 161; diaspora scholarship 6; food culture 126–132; identity/Chineseness 19–20, 40, 74, 117, 121, 124–127, 133, 140, 144, 146–150, 154–161, 163–166, 174–175, 178; language, *see* Chinese language; medicine 1; migrant females in Australia, 1911–1961 63–64, **64**; migration and settlement in Australian context 12–17; patriarchy 7–9, 71, 88–89, 91–95, 97–98, 100, 102, 110,

136, 140, 163; restaurants and cafes 127, 129; students 96; wives 7–8, 12

Chinese Australian: business 116; cultural maintenance/abandonment 115–117; families and social isolation 116–117; immigration 16; men 5–6, 9, 26, 29, 88; in neighbourhoods and communities 117; scholarship 5–6, 61, 169–170, 173

Chinese Australian community 5, 157; difference and Otherness 150; female contributions to 10–11, 74, 87–88, 91, 99–112; sense of isolation 116–117; in Sydney 135

Chinese Australian females: age groups, 1911–1966 61, **62**; assumptions of foreignness 149–153; birthplace of 53–56, **53–55**; census records 9–10, 39–40, 56–57, 61, 177; country of birth 54, **55**; cultural maintenance 19, 174–175; dependants in Australia, 1911 106, **107**; in diaspora 17; domestic roles, *see* homes and family responsibilities; education 18, 90, 93–99, 110, 112n1, 112n7, 116, 118, 120, 153, 155, 165–167, 174–175; employment; experiences 89, 102–110, 131, 144, 153, 155, 166–167, 172–175; statistics 106, **108–109**; erasure of 7; everyday' discriminatory practices and prejudice 153–156; experiences of inclusion 161–166; in family businesses 19, 99–110; family decisions on mobility 81–82; gender inequality 8, 91, 95, 174; geographical location 77–83, **78**, **80**; homes and family economies, *see* homemaking and family economy; homes and family responsibilities 89–92, 94–105, 115–116, 118–126, 129–132, 172–174; interactions and identity beyond home 175; 'lack of' 5–6; literacy of 118, **119**; longstanding presence of 63–68; marriage 59; memories of difference 159; migration and mobility 15, 68–83, **70–71**; national and cultural identities 145; in 1911 118; occupations of 106, **108–109**; population 39–41, 51; presence 52–68, 144, 166, 171–172; in research process 37–39; rural/urban distributions, 1911–1966 79, **80**, 81; school-yard taunting and bullying 145–149; state distributions, 1911–1966 77–79, **78**; unmarried 60–61; unpaid work 19, 29, 110; voices of 35–39

Chinese diaspora 6, 8, 17, 20, 43, 54, 99, 116, 169, 178

Chinese language: Cantonese 118–119, 122, 147; dialects 125; families 127; fathers, language and identity 124–126; maintenance 118–126; mothers language and identity 119–124; newspapers 9, 116, 119, 141n2; skills 119–121

Chinese Quarter 116

Chinese rituals and seasonal festivities: ancestor worship, *Qingming* 137–139; maintenance 132–139; New Year celebrations 134–137

Choi, C. 5, 8–9, 21n9, 41, 56, 61, 69, 71, 80, 85n5, 112n5

class (socio-economic) 111, 156–159, 164

Colombo Plan 16, 54

Confucian family system 8, 18–19, 72

Confucian ideology 88–89, 92, 95, 97, 99, 102, 110–111

Confucianism 136; in Australian family contexts 88–92; in childhood, *see* childhood memories; gender expectations 94–99

Coolie migration 12–13

Cope, M. 28

Couchman, S. 11, 169

Crenshaw, K. W. 32

cultural custodian 115, 121–122, 130, 136, 138–139, 175–176

Denson, N. 167n1

domestic realm 115

Domosh, M. 29–30

double isolation 117

Dowling, R. 36, 38

Dua, E. 30

Dunn, K. 167n1

Dunn, K. M. 132

Ebrey, P. B. 89

employment, *see* Chinese Australian females, employment

English language/literacy 118–120, **119**

family business, *see* homemaking and family economy

Fay family 66, 66, 67

'Federation Census' of 1901 44n5

female absence 6–8, 27, 69

female (definition) 20n3

female presence: Australian-born female children 61, **62**, 62–63; birthplace

statistics **53**, 53–56, **55**; enduring
presence in Australia 63–68, **64**, 65–68;
full and mixed Chinese in Australia,
1901–1971 52, **52–53**; unmarried
Chinese Australian women 60–61;
wives and mothers 56–59
feminism 25
feminist geographers 30, 33, 35–36
feminist historical geography: emergence
and themes 17, 25–29, 42–43, 115,
170; methodological challenges 29–30,
43; national epic style 27–29; in North
America 25–27, 29–31
Fitzgerald, J. 5, 112n5
Fitzgerald, S. 153
Fong See Lee Yan 102, *103*
food culture and Chineseness 126–128;
Chinese cuisine 131–132; mothers and
maintenance of 128–131
foreign-born females 69

Gassin, G. 11, 169
Gates, H. 8
gender roles 3, 38, 88–89, 92, 94–95, 100,
111, 156, 172
Giese, D. 96, 169
Gittins, J. 75
Gleeson, B. 31, 176
Goffman, E. 112n4
goldfields 13, 67, 79, 150
Gold Mountain 8, 43, 72
Greer, G. 98
Gulley, H. 30
Guangdong Province 13, 72, 118, 185,
187–189, 192
Guomindang (Kuo Min Tang, KMT)
11, 116

Hage, G. 151, 159–161
Ham Hop (Mrs Poon Gooey wife of Poon
Gooey) 7, 21n4
Hanson, S. 147
Heffernan, M. 30
historical geographies 42–43; and
contemporary contexts 42–43; feminist,
see feminist historical geography;
reconstructing Chinese Australian
176–177
historical sources 9
home and household geographies 27–29,
88–105, 110–111, 115, 117, 119–121,
123–124, 128–132, 134–137, 140–141
homemaking and family economy 87,
99, 112n1; employment beyond family

businesses 105–110, **107–109**; role
in family business 102–106, *103–105*;
women's work 100–102
Hong Kong 1–2, 16, 36, 43n1, 54, **55**, 59,
61, 67, 74–77, 84n2, 97, 110, 117–118,
122, 132, 147, 161, 185–189, 191–192
Hong Yuen department store 104, *105*,
187, 191
Honig, B. 116, 175
hooks, b. 116, 175
Howitt, R. 41
Huck, A. 96, 112n5
Hudson, B. 35
Huggins, J. 44n4

identity: Australian 19–20, 42, 118, 126,
144–149, 151–152, 159–161, 166,
175–176; Chinese 19–20, 40, 74, 117,
121, 124–127, 133, 140, 144, 146–151,
154–161, 163–166, 174–175, 178;
Chinese Australian 10–11, 15, 17–18,
26, 34, 144–145, 170–171, 174–175;
intersectional 10, 18; national 20,
41–42, 145, 147–148, 150–151, 159,
161, 170, 177–178; negotiation 10, 31,
96, 140; place-based 27, 31–32, 34, 42,
129, 147, 176
Immigration Restriction Act 1901 3, 14,
51, 145
immigration restrictions 3, 5, 7, 9, 12, 14,
18, 51, 57, 72–73, 84, 96
Inglis, C. 110, 117
international migration and mobility:
Australian-born females 75–77;
Chinese female entry to Australia
15; Chinese migrant arrivals 69,
70–71; geographical distribution and
intra-national mobility 77–83; male
migration 68; migrant wives 71–75
intersectional approach 18, 26, 32–35
intersectionality 4, 10, 17–18, 20, 25–44,
155–156
interview: with Chinese Australian
women, *see* interview participants;
experience 38; insider/outsider status
38–39; power relationship 37–38
interview participants: Daphne Lowe
Kelley 187; Doreen Cheong 189–190;
Edith 191; Eileen Yip 191–192;
Helen Sham-Ho 192–193; Ina 185;
information **183–184**; Kaylin Simpson
Lee 190; Lily Xiao Hong Lee 186;
Mabel Lee 185; Margaret 192; Marina
Mar 190–191; Mary Tang 189; Mei Lin

Yum 186–187; Nancy Buggy 188–189; Patricia Yip 192; Sally Pang 188; Sandy 185; Stella Sun 187–188; Susan 186
intra-national mobility 77–83
Inverell 75, 81, 103, *105*, **184**, 190
Ip, D. 132
Ip, M. 9, 76, 172

Johnson, J. T. 41
Johnson, L. 35
Jones, P. 78, 85n7

Kamp, A. 21n6, 43n1, 84n2, 112n1, 167n1; interviews with women, *see* interview participants
Kay, J. 25–29
Khoo, T. 11, 169
Kobayashi, A. 30
Kuo, M. F. 112n5

Legg, S. 116, 175
legislation 3–4, 7, 12–14, 16, 18, 40, 51–52, 71, 74, 88, 105, 145, 171, 178
Ling, H. 9, 112n5, 172
Lin, Y. 135
Li, P. S. 105
Loh, M. 11, 29
London, H. I. 21n9
Loong family 67, 68
Lumbiew family 64, *65*

Macgregor, P. 51, 56, 57
marriage 1, 8, 30, 34, 36, 39–40, 56, 57, 60, 71–72, 74–76, 81–83, 90, 99, 172, 177, 187–188, 191–192; endogamy 59, 61, 76; marriage status of 'full' and 'mixed' Chinese females in Australia, 1911–1961 **58**; reunification of husbands and wives 6, 13
Martínez, J. T. 112n5
Mays, V. M. 32
McDowell, L. 37
McGurty, E. 29
McKeown, A. 8
McKewon, E. 31, 176
Medlicot, C. 30
Melbourne 11, 31, 67, 68, 83, 96–97, 103, 116, 185–186, 188
Merrylands, Sydney 61, 63
migrant mothers 64, 68, 106, 110, 120–121, 128, 139, 174
migrant wives 5–9, 12–15, 51, 56, 57, 61, 71–75, 83, 106, 153, 171–172, 177
migration literature 6–7, 26, 31, 57, 105, 171

migration studies, scholarship 6, 61, 169–170
mixed race 31, 38, 107
Modern China café 103, 128, 135, 188
modern geography 35
Mohammad, R. 33–34
Morin, K. 30
Morin, K. M. 29–30, 42
multiculturalism 11, 42, *158*, 176–178

Nast, H. J. 38
nationalised immigration restriction 14
nationality 160
naturalisation 9, 15–16, 56, 72–74, 153
New South Wales (NSW) 77–78, **78**, **80**, 81–82
New Year's Eve 133–134
New Zealand Chinese women 9, 11
non-European immigration 3
Noonan, R. 11, 169
North American historical geographies 25–26; androcentrism of 27; inclusion of women 29
Northern Territory (NT) 78–79, **78**, **80**

Ogborn, M. 25–26, 29, 176
orientalism, orientalist views 10, 19, 88, 91, 98, 111, 116, 141, 159, 171, 173–174

Parramatta, Sydney 93, 101–102, 185–186; patriarchy, experiences of 7–9, 71, 88–89, 91–95, 97–98, 100, 102, 110–111, 136, 141, 163
Peake, L. 30
Peters, E. 41
Pile, S. 37
Poon Gooey case 7
population 5, 9, 12–14, 16, 18, 39–41, 44n5, 51–54, 56, 57–61, 63–64, 69, 77–80, 83, 85n7, 106, 119, 141n2, 158, 170–171, 198–199
postcolonial/postcolonial feminism 17, 19–20, 26–27, 38–43, 129, 158, 161, 175–176; intersectional approaches in geography 32–35
postcolonial feminist approach 10–11, 17, 19, 26, 32–42, 51, 110, 115, 140, 158, 166, 169–170, 178
postcolonial feminist geography 25–44, 87, 170
Pratt, G. 147
presence: in Australia, 1911–1961 40–41, **53**, 54, **55**, **58**, 61, **62**, 63–64, **64**, 77–79, **78**, **80**; length of female Chinese migrant residence 39, 63, **64**, 170

public/private dichotomy 28, 105
Pulsipher, L. M. 30

Queensland (QLD) 77–78, **78, 80**, 81

racialisation 17, 40, 145, 149–150,
    152–153, 166, 175
racialised census data 40–41
racism 4, 15, 32, 35, 145, 149, 153–155,
    164, 167n1, 171
Ramsay, G. 31
Ramsey, C. 96
Ratnam, C. 178
Rose, G. 25–26, 29, 176, 178
Royal Geographical Society 41
Ruby Wong Chee 66, 75–76
Ryan, J. 6

Said, Edward 32
Salmon, Carty, M. P. 14
Schein, R. H. 42
school 1–2, 4, 19, 36, 39, 60–61, 67, 77,
    81–82, 90, 92–94, 96–97, 99, 101, 106,
    107, 110, 112n6, 116–117, 121–124,
    132, 144–150, 154–156, 157, *158*, 160,
    162–163, 165–166, 175, **183–184**,
    185–192
Schuurman, N. 30–31
self-reflexivity 37
Sharples, R. 167n1
Sisko, S. 167n1
Smith, K. E. 37
Song, M. 105, 112n4
Spender, D. 98
Spivak, G. C. 32
subaltern groups 35–36, 41,
    43, 170
Summers, A. 98
Sydney 3–4, 12, 31, 57–61, 63, 64, 65, 66,
    72, 74–75, 81–82, 90, 92, 93, 99–101,
    103, *104*, 117, 127, 132–133, 135,
    156–157, **183–184**, 185–192; suburbs
    81, 98, 116, 128, 151, 162, 185, 188,
    190, 192

Tan, C. 145–146, 148–150, 152–154,
    165, 169
Tart Lumbiew and Agnes Mary 64, 65
Teather, E. K. 31, 176
Teng, J. E. 88
Teo, S. E. 117

Texas, Queensland 3, 13, 61, 62, 77,
    81–82, 85n7, 133, *134*, **183**, 187
Thursday Island 1–4, 10, 76, 81–82, 131,
    184, 187–188
Tolia-Kelly, D. P. 178
Tomlinson, B. 32
traditional Chinese culture 19, 98, 178;
    family system 111

University 36, 57, 74, 84n2, 93, 96–97,
    99–100, 112n6, 141n1, 157, 178,
    185–192

Vergani, M. 167n1
Victoria (VIC) 77–78, **78, 80**

Walton, J. 167n1
Wang, G. 8
Watson, R. S. 89
Western thinking 92
White (definition) 20n2
White Australia Policy 3; context 159,
    163; objectives of 14; rationale of 56
White Australia/White Australia Policy
    era 4–5, 10, 15, 17, 36, 40, 51, 60–61,
    77, 83–84, 87–89, 99–100, 102, 105,
    110–111, 115–116, 118, 126, 144–145,
    152, 157, 159, 165–167, 169–171, 173,
    176–179
Whitlam Labor government 14
White racial imperialism 32, 44n4
White settler society 30; national
    development of 31
White society 115
Williams, M. 6
Wilton, J. 75, 105–106, 118
Wolf, M. 89
World War II 3, 51, 53, 64, 71, 73, 82, 90,
    141n2, 192
women's voices 35–39

Yamamoto, T. 150
Yang, W. 150
Yarwood, A. 6, 9
Yit Quay Loong and Yuen Mu Loong
    67, 68
Yong, C. 6–7
Young, I. M. 116, 175
Yu, H. 150
Yung, J. 9, 172

Zagumny, L. L. 30

Printed in the United States
by Baker & Taylor Publisher Services